Engineering Materials

This series provides topical information on innovative, structural and functional materials and composites with applications in optical, electrical, mechanical, civil, aeronautical, medical, bio- and nano-engineering. The individual volumes are complete, comprehensive monographs covering the structure, properties, manufacturing process and applications of these materials. This multidisciplinary series is devoted to professionals, students and all those interested in the latest developments in the Materials Science field, that look for a carefully selected collection of high quality review articles on their respective field of expertise.

Indexed at Compendex (2021) and Scopus (2022)

Carlton A. Taft • Sergio Ricardo de Lazaro
Editors

Innovative Materials for Energy and Sustainability

Springer

Editors
Carlton A. Taft
Centro Brasileiro de Pesquisas Físicas
Rio de Janeiro, Rio de Janeiro, Brazil

Sergio Ricardo de Lazaro
UFPG
State University of Ponta Grossa
Ponta Grossa, Paraná, Brazil

ISSN 1612-1317 ISSN 1868-1212 (electronic)
Engineering Materials
ISBN 978-3-032-09148-2 ISBN 978-3-032-09149-9 (eBook)
https://doi.org/10.1007/978-3-032-09149-9

This Springer imprint is published by the registered company Springer Nature Switzerland AG
The registered company address is: Gewerbestrasse 11, 6330 Cham, Switzerland

Contents

A Comprehensive Look at MoS_2 Nanostructures: From Synthesis to Tailored Functionality

Felipe Moreira Pinto, Fernando Wypych, Carlton A. Taft, and Felipe de Almeida La Porta

Abstract This chapter delves deep into the latest advancements in synthesizing and modifying molybdenum disulfide (MoS_2) nanoparticles with tailored morphologies, ranging from zero-dimensional to three-dimensional, which exhibit distinct physicochemical properties and multifunctionality. We discuss key chemical synthesis routes—including hydrothermal and solvothermal methods, thermal decomposition of precursors, and chemical vapor deposition—are discussed, with a focus on their benefits for controlling morphology and phase composition. Physical methods like mechanical milling and various exfoliation techniques, such as ultrasonication-assisted liquid exfoliation and electrochemical exfoliation, are also explored for their effectiveness in producing few-layer or monolayer MoS_2 nanoparticles. Furthermore, the review delves into crucial surface modification strategies, including metal and non-metallic doping, as well as defect engineering (specifically sulfur vacancies and edge functionalization) and phase transition control (2H to 1T

F. M. Pinto
Federal Institute of Education, Science and Technology of Minas Gerais, Bambuí Campus, Bambuí, MG, Brazil

F. Wypych
Post-Graduation Program in Materials Science and Engineering, Federal University of Technology, Londrina, PR, Brazil

Department of Chemistry, Federal University of Paraná, Curitiba, PR, Brazil

C. A. Taft
Brazilian Center for Research in Physics, Rio de Janeiro, Brazil

F. A. La Porta (✉)
Post-Graduation Program in Materials Science and Engineering, Federal University of Technology, Londrina, PR, Brazil

Post-Graduation Program in Chemistry, State University of Londrina, Londrina, Paraná, Brazil

Faculty of Informatics and Management, University of Hradec Kralove, Hradec Kralove, Czech Republic
e-mail: felipelaporta@utfpr.edu.br

C. A. Taft, S. R. d. Lazaro (eds.), *Innovative Materials for Energy and Sustainability*, Engineering Materials,
https://doi.org/10.1007/978-3-032-09149-9_1

conversion). These modifications are pivotal for enhancing electronic conductivity, catalytic efficiency, and overall functional performance, enabling MoS_2-based innovations in nanoelectronics, electrocatalysis, energy storage, and biomedicine. This work aims to serve as a strategic guide for the rational design of MoS_2 nanomaterials with tailored morphologies for next-generation advanced functional devices.

Keywords MoS_2 nanostructures · Synthesis methods · Surface modification · Defect engineering · Exfoliation techniques

1 Introduction

Nanotechnology, as we know it, began with a profound idea articulated by Richard Feynman's visionary 1959 lecture, "*There's Plenty of Room at the Bottom.*" In this talk, Feynman introduced the concept of directly manipulating matter at the atomic scale, thereby laying the conceptual groundwork for the field. Therefore, from this perspective, nanotechnology involves engineering advanced nanomaterials—where at least one dimension falls between 1 and 100 nm—possessing unique physicochemical properties fundamentally different from their bulk counterparts owing to quantum and surface-area effects at the nanoscale [1–3].

Among advanced nanomaterials, molybdenum disulfide (MoS_2)—as a leading two-dimensional (2D) transition metal dichalcogenide (TMDC)—exhibits a unique electronic structure and remarkable optical and catalytic nanoscale properties, which attract widespread scientific interest, especially for advanced applications of nanotechnology [4, 5]. Its graphene-like layered structure provides highly tunable physicochemical characteristics, most notably a layer-dependent intrinsic bandgap—a key feature absent in graphene—which enables its use in semiconducting and optoelectronic devices [6]. Additionally, its high mechanical flexibility, chemical stability, and the presence of multiple polymorphic and polytypic phases (1T, including distorted variants such as 1T′, 1T″ and 1T-‴, along with 2H, and 3R) further expand its applicability in energy storage, catalysis, and sensing technologies [7, 8]. Notably, the 2H phase is the most thermodynamically stable polymorph/polytype, featuring a semiconducting trigonal prismatic structure, while the 1T phase is a metastable, metallic polymorph with an octahedral crystal structure [5]. Although metastable, the 1T phase has higher electronic conductivity and more active sites for catalysis, advantages that are particularly pronounced in 1T/2H heterostructures [9–11]. The 3R phase of MoS_2, a rhombohedral polymorph/polytype, has recently gained attention for its superior performance in hydrogen evolution reactions (HER) compared to the 2H phase, and it also facilitates easier conversion to the highly active 1T phase, making it a promising precursor for diverse catalytic applications [5, 12, 13]. Additionally, it is well-known that the synergistic interplay between metallic and semiconducting domains, combined with the creation of mid-gap states in 1T/2H and 1T/3R hybrids, is key to their enhanced photocatalytic and piezo-catalytic activities. Hence, this mechanism fundamentally improves charge carrier separation and transport, thereby amplifying catalytic activity [14, 15].

In addition to its intrinsic electronic and phase characteristics, a nanomaterial's morphology—its shape and dimensionality—is a key determinant of its final properties and function. Consequently, significant efforts have recently been focused on developing methodologies for the controlled synthesis of MoS_2-based nanomaterials, with a strong emphasis on precisely regulating their particle size and morphology to tailor their resulting functionalities [5, 16–19]. Such synthetic approaches have been extended to create hierarchical design of MoS_2 nanostructures with different morphologies—zero-dimensional (0D) to three-dimensional (3D)—which offering unique physicochemical properties and multifunctionality [20–22]. These are commonly synthesized via solvothermal methods or spray pyrolysis, where controlled aggregation of MoS_2 precursors leads to spherical structures [22–24]. By adjusting experimental parameters of the reaction, e.g., such as precursor concentration, reaction temperature, and solvent composition [25], researchers can precisely control the size, porosity, and surface roughness of the nanospheres [26]. The one-dimensional (1D) MoS_2 nanotubes or nanorods have also received considerable attention owing to their high aspect ratios, superior mechanical strength, and excellent charge transport properties [27, 28]. As we know, these characteristics make them ideal for energy storage devices, such as supercapacitors and lithium-ion batteries, and for tribological applications where high wear resistance is essential [20]. Particularly, the synthesis of MoS_2 nanotubes typically involves either template-assisted growth or self-assembly. In template-assisted methods, scaffolds like carbon nanotubes or anodic aluminum oxide guide the conformal deposition of MoS_2 layers [29]. Alternatively, high-temperature self-assembly via vapor-phase or solvothermal reactions can facilitate their spontaneous formation [27].

Recent studies have advanced this concept by creating hierarchical structures, such as nanospheres composed of petal-like nanoparticles. These hybrid morphologies combine the benefits of both shapes, offering increased active edge sites and improved charge transport for superior catalytic performance [30]. In parallel, surface modification strategies—including doping, functionalization, and defect engineering—are employed to enhance electronic conductivity [31], catalytic efficiency [32, 33], and integration into composite materials [34, 35]. These combined advancements are driving MoS_2-based innovations in nanoelectronics, electrocatalysis, energy storage, biomedical applications, environmental remediation, and so on [36, 37].

Hence, the objective of this chapter is to serve as a strategic guide for the rational design of MoS_2 nanomaterials with tailored morphologies for next-generation advanced functional devices. The following sections will delve into the latest advancements in synthesizing and modifying MoS_2 nanostructures, which exhibit distinct physicochemical properties and multifunctionality. We will discuss key chemical synthesis routes (such as hydrothermal and solvothermal methods, thermal decomposition of precursors, and chemical vapor deposition (CVD)), with a focus on their benefits for controlling morphology and phase composition. Physical methods like mechanical milling and various exfoliation techniques, such as ultrasonication-assisted liquid exfoliation and electrochemical exfoliation, will also be explored for their effectiveness in producing few-layer or monolayer MoS_2 nanoparticles. Furthermore, this work will delve into crucial surface modification strategies, including metal and non-metallic doping, as well as defect engineering (specifically sulfur vacancies and edge functionalization) and phase transition control (2H to 1T conversion). These modifications are pivotal for enhancing electronic conductivity, catalytic efficiency, and overall functional performance, enabling MoS_2-based innovations in nanoelectronics, electrocatalysis, energy storage, and biomedicine.

2 Advanced Methodologies for MoS_2-Based Nanomaterials Synthesis with Tailored Morphologies

The synthesis of MoS_2 nanomaterials with well-defined morphologies—such as nanoparticles, nanotubes, nanoflowers, nanospheres, and so on—have attracted significant interest owing to their tunable physicochemical properties and extensive application potential [7, 8]. The primary synthesis approaches can be categorized into chemical and physical methods, along with exfoliation strategies for obtaining layered MoS_2 nanoparticles [16, 38]. These methodologies are crucial for the rational hierarchical design of MoS_2 nanostructures for next-generation functional devices, and will be discussed in detail below.

2.1 Chemical Synthesis of MoS_2-Based Nanomaterials

As we know, the chemical synthesis routes (i.e., which build advanced materials from molecular precursors) are highly valued for their potential for scalability and cost-effectiveness, as well as their ability to yield nanostructures with well-controlled morphology and phase composition [39]. A key advantage of chemical synthesis over physical methods is the ability to control stoichiometry, defect engineering, and doping precisely [40–44]. Among the main chemical strategies for MoS_2-based nanomaterial synthesis are hydrothermal and solvothermal methods, thermal decomposition of precursors, and the CVD.

2.1.1 Hydrothermal and Solvothermal Synthesis

Hydrothermal and solvothermal synthesis are highly effective solution-based methods that rely on liquid-phase reactions to control the nucleation and growth of MoS_2 with high crystallinity [45]. Their primary advantages include low-temperature operation, ensuring both high purity and energy efficiency, and a high degree of control over structural parameters (e.g., such as layer number, crystallite size, and/or more or less defects) [46]. Such structural control is fundamental to optimizing MoS_2 for targeted functional applications [47, 48]. These solution-based techniques involve reacting molybdenum and sulfur precursors in water (hydrothermal) or organic solvents (solvothermal) under elevated temperature and pressure, enabling the formation of diverse MoS_2-based nanostructures, such as nanoparticles, nanoflowers, and quantum dots [47, 49–52].

Notably, the morphology of MoS_2-based nanomaterials produced via hydrothermal/solvothermal synthesis can be tailored by adjusting reaction conditions. The hydrothermal synthesis of MoS_2 involves a chemical reaction between molybdenum precursors and sulfur source in an aqueous medium (Reaction 1). Common molybdenum precursors include ammonium tetrathiomolybdate ($(NH_4)_2MoS_4$) or sodium molybdate (Na_2MoO_4), while typical sulfur sources such as thiourea (CH_4N_2S), thioacetamide (C_2H_5NS), or elemental sulfur (S_8). This reaction takes place in a sealed autoclave, usually at temperatures between 150 and 300 °C. The kinetics of this reaction and the final material properties are critically dependent on the experimental parameters (e.g., temperature, reaction time, pH, surfactants or stabilizing agents, and so on) [53].

$$(NH_4)_2\,MoS_4 \rightarrow MoS_2 + 2NH_3 + H_2S \quad (1)$$

At lower reaction temperatures, typically below 180 °C, the resulting MoS_2 is often amorphous or poorly crystalline. However, increasing the temperature to 230–260 °C promotes high crystallinity and favors the formation of desirable monolayer or few-layer nanoparticles. At temperatures exceeding 300 °C, excessive growth and aggregation of MoS_2 nanoparticles, resulting in larger polyhedral structures with reduced catalytically active surface area [54]. On the other hand, however, short reaction times (<6 h) result in incomplete crystallization and the formation of ultrathin MoS_2 nanoparticles and frequently contaminated with MoS_3. In contrast, prolonged reaction durations, ranging from 24 to 48 h, promote the growth of well-defined, layered structures with enhanced crystallinity and structural stability. Similarly, in acidic conditions (pH <6), MoS_2 tends to form nanospheres owing to the rapid nucleation, whereas near-neutral pH conditions are the optimal for synthesizing monolayer. Acids conditions (pH <6), however, can lead to the formation of unwanted MoO_3 impurities [55].

The introduction of surfactants and structure-directing agents further refines the morphology and functional properties of hydrothermally synthesized MoS_2. Commonly used surfactants, including the non-ionic polyvinylpyrrolidone (PVP), the cationic cetyltrimethylammonium bromide (CTAB), as well as the anionic sodium dodecyl sulfate (SDS), are effective in regulating nanoparticle dispersion and preventing agglomeration, which in turn leads to diverse morphologies [56–58]. For instance, the synthesis of MoS_2 nanotubes requires the presence of capping agents such as ethylene glycol, which promote the rolling of nanoparticles into tubular structures [59]. The polarity of the solvent is a critical parameter, as it directly alters precursor solubility and, consequently, the reaction kinetics. For example, dimethylformamide (DMF) functions not only as a solvent but also as a reducing agent that facilitates the Mo^{6+} to Mo^{4+} reduction to yield high-purity MoS_2. In contrast, other solvents (e.g., ethanol and ethylene glycol) are usually chosen to produce smaller, more monodisperse nanoparticles with controlled surface chemistry, which is particularly advantageous for catalytic and biomedical applications [60]. Hydrothermal and solvothermal synthesis methods offer an unparalleled platform for the rational design and fabrication of MoS_2-based nanomaterials with controlled morphologies. Future research will likely focus on developing continuous-flow hydrothermal reactors for large-scale synthesis and on using advanced computational modeling to better predict how synthesis parameters translate to specific morphologies [46].

2.1.2 Thermal Decomposition of Precursors

Thermal decomposition of molybdenum-containing precursors is a highly efficient and scalable strategy for producing pure, crystalline MoS_2 nanostructures. Based on the pyrolytic breakdown of molybdenum-containing compounds in a controlled atmosphere, thermal decomposition is a dry-process method that offers enhanced control over the final MoS_2 nanostructure and phase composition compared to wet-chemical routes. The final properties—including crystallite size, morphology, and defect concentration—are precisely tuned by the careful selection of precursors and the optimization of reaction conditions [61, 62].

The thermal decomposition synthesis of MoS_2 nanostructures utilizes variety of precursors, including tetrabutylammonium molybdate ($((C_4H_9)_4N)_2MoO_4$, ammonium tetrathiomolybdate ($(NH_4)_2MoS_4$), molybdenocene dichloride ($Mo(C_5H_5)_2Cl_2$), and molybdenum hexacarbonyl ($Mo(CO)_6$), which decompose upon heating to form the desired nanostructures. Of these, ammonium tetrathiomolybdate is particularly advantageous as a single-source precursor because it already contains both molybdenum and sulfur elements, thereby eliminating the need for additional sulfurizing agents [63, 64].

The thermal decomposition process follows distinct reaction pathways depending on the precursor (e.g., single-source or multi-source) [63]. In single-source systems, such as the decomposition of $(NH_4)_2MoS_4$ is a well-defined, stepwise process. At lower temperatures, the precursor breaks down to form an amorphous, sulfur-rich intermediate, molybdenum trisulfide (MoS_3), while releasing ammonia (NH_3) and hydrogen sulfide (H_2S) gas (Reaction 2). As the temperature increases further, this MoS_3 intermediate is subsequently reduced to form crystalline MoS_2 by releasing elemental sulfur (Reaction 3) [65, 66].

$$\left(NH_4\right)_2 MoS_4 \rightarrow MoS_3 + H_2S + 2NH_3 \tag{2}$$

$$MoS_3 \rightarrow MoS_2 + S \tag{3}$$

However, the presence of an inert gas atmosphere is critical for preventing unwanted oxidation (i.e., the formation of molybdenum oxides) and promoting the formation of stoichiometric, high-purity MoS_2. In some cases, H_2 or H_2S gas is introduced to create a reducing environment that further promotes the formation of high-purity MoS_2 with minimal oxide contamination [65].

Alternatively, in multi-source systems, such as the decomposition of molybdenum hexacarbonyl occurs via the dissociation of CO ligands, which are released as gaseous byproducts, leading to the nucleation of MoS_2 nanostructures in the presence of a sulfur source (Reaction 4) [67]:

$$4\text{Mo}(\text{CO})_6 + \text{S}_8 \rightarrow 4\text{MoS}_2 + 24\text{CO} \tag{4}$$

The final properties of MoS_2 synthesized via thermal decomposition are critically dependent on the thermal parameters such as temperature, heating rate, and dwell time. Lower temperatures (typically below 100 °C), favor the formation of the metastable, metallic 1T phase, which is highly desired for energy storage and electrocatalysis. At higher temperatures, generally exceeding 600 °C, the thermodynamically stable 2H phase is preferentially formed, characterized by its semiconducting nature and suitability for electronic and optoelectronic applications [65, 67].

The 1T-to-2H phase transition is largely governed by the kinetics of sulfur vacancy formation and atomic rearrangement. Studies have shown that the introduction of alkali metal intercalants, such as lithium or sodium, during the decomposition process can stabilize the 1T phase by preventing structural relaxation into the more energetically favored 2H configuration. This has led to the development of lithium-intercalated MoS_2 materials, which demonstrate enhanced performance in energy applications, including the HER and supercapacitors [66]. Furthermore, by adjusting the parameters such as precursor concentration and the gas flow rate, the morphology and crystallite size can be tuned, enabling the synthesis of diverse MoS_2 nanostructures, including ultrathin nanoparticles, nanospheres, and hierarchical flower-like assemblies [61–63, 65–67] .

In the first synthesis of 1T-MoS_2, potassium tetrathiomolybdate (K_2MoS_4) was synthesized by sulfidation of potassium molybdate, followed by reduction under a N_2/H_2 atmosphere at 850 °C for 72 h to yield $KMoS_2$. Subsequent washing with distilled water produced the hydrated phase $K_x(H_2O)_yMoS_2$, during which partial oxidation and KOH formation occurred. The 1T-MoS_2 phase was obtained through three distinct methods: (1) chemical oxidation using $K_2Cr_2O_7$ in acidic medium, (2) galvanostatic oxidation in acetonitrile, and (3) chemical oxidation with iodine in acetonitrile under argon. All methods resulted in golden-lustered 1T-MoS_2 crystals, preserving the octahedral coordination of Mo and yielding high-quality single crystals up to several millimeters—surpassing typical results from conventional synthesis routes [68, 69].

2.1.3 Chemical Vapor Deposition

CVD has emerged as a premier technique for hierarchical design of high-quality MoS_2 nanostructures with precise control over layer number, phase purity, and crystallinity [70–74]. This technique is based on the vapor-phase reaction of molybdenum and sulfur precursors on a heated substrate, where controlled nucleation and growth enable the formation of monolayer and few-layer MoS_2 films [75]. The general reaction mechanism involves the reduction and sulfurization of a molybdenum precursor, such as molybdenum trioxide (MoO_3), by a sulfurizing agent, typically sulfur powder (S_8) or H_2S [76]. The success of CVD synthesis is determined by a set of critical process parameters. Synthesis temperatures typically range between 600 and 900 °C. This thermal window critically influences both nucleation density and crystal growth kinetics, affecting the final number of layers; for instance, higher temperatures often provide the necessary energy for the growth of large, high-quality monolayer domains, provided that precursor flux is optimized [77, 78]. The carrier gas, often argon or a mixture of argon and hydrogen, is crucial for controlling precursors' transport and deposition kinetics. Also, it is well-known that the substrate significantly influences MoS_2 nucleation density and film uniformity, with materials like SiO_2/Si, sapphire, and gold-coated silicon wafers commonly employed to promote epitaxial growth, leading to highly oriented films with minimal domain boundaries [78, 79]. By elucidating the fundamentals of nucleation, gas-phase and surface chemistry, and transport phenomena, mechanistic studies have enabled the precise control and tailoring of growth processes at the atomic level [80].

The ability to grow material on wafer-scale substrates has positioned CVD as a leading technique for integrating TMDs into next-generation semiconductor technologies. Nevertheless, significant challenges remain, including high operational costs, long processing times, and the requirement for vacuum or controlled-atmosphere systems, which currently limit its industrial scalability [76]. To overcome these limitations, alternative CVD-based techniques such as plasma-enhanced CVD (PECVD) [81, 82] and metal-organic CVD (MOCVD) [83, 84] have been developed. PECVD enables MoS_2 growth at significantly lower temperatures, typically 200–400 °C, by utilizing plasma to activate precursor dissociation and drive the necessary surface chemical reactions [85]. This low-temperature capability is a key advantage of PECVD, as it enables the deposition of high-quality thin films on a wide variety of thermally sensitive substrates—including polymers, plastics, and fully fabricated semiconductor wafers with metallic interconnects—without causing damage or degradation [85, 86]. For instance, crystalline MoS_2 thin films have been successfully synthesized directly on flexible plastic substrates at temperatures below 300 °C, a critical step for developing next-generation wearable and transparent electronics [86]. Despite its advantages, translating PECVD to commercial applications faces significant hurdles in achieving large-area uniformity, industrial-scale throughput, and precise control over material properties [87]. In parallel with PECVD's focus on low-temperature growth, MOCVD has been developed to address other key challenges, primarily achieving superior film uniformity and precise compositional control. This technique employs metal-organic precursors that enable highly reproducible deposition over large areas [88–90]. A unique advantage of MOCVD is its capability for in-situ doping, which is essential for creating complementary logic circuits (i.e., both n-type and p-type transistors) on the same chip. While the properties of MOCVD-grown MoS_2 are sensitive to growth temperature—with higher temperatures improving crystallinity but increasing the risk of sulfur vacancies—the resulting high quality and uniformity make it the method of choice for high-performance electronic devices [91]. Additionally, this has enabled the fabrication of large arrays of field-effect transistors (FETs) with consistent performance characteristics [92]. Furthermore, advanced strategies applicable to both MOCVD and other CVD techniques, such as substrate engineering, have enabled the selective growth of MoS_2 on pre-patterned nucleation sites. This allows for the direct fabrication of heterostructures and complex device architectures [93]. While these advanced methods offer remarkable control, further research is needed to develop cost-effective, scalable synthesis routes that maintain the exceptional quality of MoS_2 films while reducing processing complexity [94].

2.2 *Physical Synthesis of MoS_2-Based Nanomaterials*

Physical methods have become a robust pathway for producing MoS_2 nanostructures at scale. Unlike chemical routes that build materials from atomic precursors, these techniques rely on applying external energy—such as mechanical or electrical force—to break down bulk MoS_2 crystals into nanoscale forms. This atomic/molecular method leverages the inherent anisotropy of MoS_2 bonding. The strong in-plane covalent bonds coupled with weak out-of-plane van der Waals forces enable the delamination of the material into multilayered structures ultimately allowing for exfoliation into monolayer films [5]. These methods are particularly valued for their cost-effectiveness, simplicity, and suitability for large-scale production without the need for complex chemical reactions. In this chapter, we will explore the key physical techniques used for this purpose, focusing on mechanical milling and various exfoliation-based strategies for producing high-quality nanoparticles.

2.2.1 Mechanical Milling

High-energy ball milling, a form of mechanical milling, has emerged as a practical method for producing MoS_2 nanoparticles by exploiting shear and impact forces to induce particle fragmentation and exfoliation [95]. The process typically involves loading bulk MoS_2 powder into a milling chamber containing hardened steel or tungsten carbide balls, which impart high mechanical energy to the material as the chamber rotates at high speeds. The repeated collision and compression of MoS_2 particles lead to structural deformation, grain refinement, and defect formation, ultimately facilitating the breakdown of bulk crystals into nanoscale structures. This method has been widely utilized for applications where defect-rich MoS_2 is advantageous, particularly in electrocatalysis and tribology, where sulfur vacancies and edge sites are crucial in boosting catalytic activity [96].

The properties of mechanically milled MoS_2 are highly dependent on key processing parameters, including milling time, ball-to-powder ratio, rotation speed, and atmospheric conditions. Extended milling durations lead to progressive particle size and thickness reduction; however, excessive milling often results in amorphization and undesired oxidation of MoS_2, normally producing oxisulfides with various degrees of sulfur substituitions represented by the general formula $MoS_{2-x}O_x$. To mitigate these effects, milling is typically conducted in an inert atmosphere, such as argon or nitrogen, which prevents oxidation and maintains phase stability. Studies have demonstrated that short-duration milling (up to 6 h) results in layered MoS_2 structures with minimal defects, while prolonged milling times (>24 h) promote the formation of highly defective and amorphous MoS_2, which may adversely affect its electronic/optical properties [97].

One of the principal limitations of mechanical milling is its broad particle size distribution, which necessitates additional processing steps to achieve uniformity. Post-milling separation techniques, including centrifugation and filtration, are often employed to isolate exfoliated nanoparticles from unprocessed bulk material. Additionally, surfactant-assisted milling has been explored to improve exfoliation efficiency and prevent nanoparticle restacking. The introduction of dispersing agents such as PVP or CTAB has been shown to enhance the stability of exfoliated MoS_2 layers in suspension, thereby increasing yield and improving processing efficiency [96]. Despite its inherent limitations, mechanical milling remains a scalable and economically viable method for synthesizing MoS_2 nanoparticles, particularly in applications where controlled defect density and high surface area are advantageous.

2.3 *Exfoliation-Based Methods*

Exfoliation-based techniques have garnered significant attention as one of the most efficient and versatile routes for isolating few-layer and single layers MoS_2 from their bulk form [98]. The inherent anisotropic bonding nature of MoS_2, as mentioned, facilitates the mechanical or chemical delamination of the layered particles under suitable external stimuli, enabling the production of atomically thin particles with minimal disruption to their crystallographic integrity.

The appeal of exfoliation methods lies in their ability to produce nanoparticles with high surface-to-volume ratios and tunable thicknesses, characteristics that are pivotal for enhancing the material's electronic, optical, and catalytic properties [99]. Moreover, exfoliation strategies provide a relatively straightforward and cost-effective means of accessing dispersions of large-area 2D MoS_2, making them particularly attractive for scalable applications in flexible electronics, energy storage systems, optoelectronics, and heterogeneous catalysis [98].

Among the most widely employed exfoliation techniques are ultrasonication-assisted liquid exfoliation, electrochemical exfoliation, and solvent-assisted exfoliation. Ultrasonication-assisted methods utilize acoustic cavitation in a liquid medium to induce shear forces that overcome interlayer interactions, whereas electrochemical exfoliation relies on redox-driven intercalation of ions to weaken van der Waals bonding, promoting the disassembly of layers. Solvent-assisted exfoliation is predicated on the dispersion of bulk MoS_2 in solvents with surface energies closely matched to that of the material, thereby enhancing the thermodynamic stability of the exfoliated single layers and reducing the tendency for restacking [100–102].

2.3.1 Ultrasonication-Assisted Liquid Exfoliation

Ultrasonication-assisted liquid exfoliation is highly accessible and widely employed physical methods for delaminating layered MoS_2 nanostructures into few-layer or monolayers [103]. This technique operates through the application of high-frequency ultrasonic waves, which generate acoustic cavitation within a liquid medium [104]. The rapid formation and collapse of microbubbles produce intense localized shear forces and shock waves capable of overcoming the weak van der Waals interactions holding MoS_2 layers in the bulk crystal structure [105]. When MoS_2 is dispersed in a suitable solvent, these forces can effectively induce exfoliation while preserving the in-plane covalent network of the single layers.

The efficacy of this method is critically dependent on the physicochemical properties of the solvent, particularly its surface energy. Solvents with surface energies closely matched to that of MoS_2 minimize the enthalpy of exfoliation, thereby enhancing dispersion stability and reducing the reaggregation of exfoliated layers [45]. Commonly employed solvents include N-methyl-2-pyrrolidone (NMP), dimethyl sulfoxide (DMSO), and isopropanol, which have demonstrated effective stabilization of MoS_2 nanoparticles through solvent-surface interactions [97, 106]. The use of such solvents facilitates the formation of stable colloidal dispersions, enabling post-processing techniques such as centrifugation and filtration to isolate nanoparticles with desired lateral dimensions and thicknesses.

Despite its practical advantages—including relative simplicity, scalability, and the absence of harsh chemical treatments—ultrasonication-assisted exfoliation faces several challenges that limit its broader application. Chief among these are the typically low exfoliation yields, prolonged processing times, and the dependence on toxic or high-boiling-point solvents, which hinder environmental and economic sustainability [107]. Moreover, extended ultrasonication may lead to structural degradation or defect formation within the nanoparticles, potentially compromising their intrinsic electronic and optical properties.

In response to these limitations, modifications to the standard ultrasonication protocol have been explored, such as the inclusion of surfactants or polymers to stabilize exfoliated single layers [108], as well as the optimization of sonication parameters including frequency, power, and duration [109]. Additionally, hybrid strategies that combine ultrasonication with chemical or electrochemical pre-treatments have shown promise in improving exfoliation efficiency and product quality [110]. Nevertheless, further research is necessary to refine these processes and enable the scalable production of high-purity MoS_2 nanoparticles tailored for specific applications in nanoelectronics, sensing, and energy conversion technologies.

2.3.2 Electrochemical Exfoliation

Electrochemical exfoliation has emerged as a highly efficient and controllable method for producing MoS_2 nanoparticles. It uses an external electric field to induce layer separation in bulk MoS_2 [111, 112]. This technique exploits the inherent anisotropic bonding of MoS_2, where strong in-plane covalent interactions within adjacent layers contrast sharply with the weak out-of-plane van der Waals forces holding the layers together [113, 114]. By applying an appropriate electric potential and partial reduction of molybdenum atoms to the +3 oxidation state, cationic intercalation is facilitated, leading to structural expansion, interlayer weakening, and eventual delamination into few-layer or monolayer [115].

The electrochemical exfoliation process typically involves a two-electrode setup, where a bulk MoS_2 working electrode is immersed in an electrolyte containing mobile cations, such as lithium (Li^+), sodium (Na^+), potassium (K^+), or ammonium (NH_4^+) ions [116]. Upon application of a bias reducing voltage, partial reduction of molybdenum atoms occurs and the cations in solutions are intercalated into the MoS_2 interlayer spaces, weakening the van der Waals interactions and inducing layer expansion [117–119]. Subsequent reduction or oxidation reactions alter the local charge distribution, further destabilizing the interlayer interactions and ultimately facilitating the exfoliation of MoS_2 into the surrounding electrolyte. The effectiveness of the exfoliation process is dictated by several key parameters, including applied potential, electrolyte composition, intercalated cations ion size, energy of cations solvation and processing duration [115, 117, 120].

The efficiency of the electrochemical exfoliation process and the quality of the resultant MoS_2 are profoundly influenced by the selection of the electrolyte, intercalated cation and the stoichiometry [121]. The prevalent use of aqueous electrolytes (e.g., LiCl, Na_2SO_4, or KOH) is attributed to their high ion mobility and amenability to scalable production protocols [122]. However, the use of organic electrolytes, including ionic liquids and acetonitrile-based solutions, has been explored to improve exfoliation selectivity and minimize oxidative damage [70, 71]. The incorporation of reducing agents in the electrolyte can further aid in preserving the intrinsic structural and electronic properties of the exfoliated MoS_2 sheets [70].

Electrochemical exfoliation presents several distinct advantages over traditional liquid-phase exfoliation techniques, including notably higher yield, significantly reduced processing time, and enhanced scalability. Additionally, unlike chemical exfoliation methods that often require harsh reagents and prolonged sonication, electrochemical exfoliation provides a cleaner and more environmentally sustainable route for obtaining high-quality MoS_2 nanoparticles with tunable thickness [117]. With their high dispersibility in polar solvents, the synthesized nanoparticles are perfectly suited for diverse applications, including energy storage, various catalytic reactions, and flexible electronic devices.

2.3.3 Solvent-Assisted Exfoliation

Solvent-assisted exfoliation constitutes a widely explored method for the delamination of layered TMDCs, particularly MoS_2, into ultrathin 2D nanoparticles [123]. This method is predicated on thermodynamic principles governing interfacial interactions between the solvent and the layered material [45]. Specifically, the efficacy of exfoliation is enhanced when the surface energy of the solvent is closely matched to that of the MoS_2 layers, thereby minimizing the Gibbs free energy change associated with the separation process. In such systems, the solvent molecules effectively intercalate into the interlayer galleries, mitigating van der Waals forces and stabilizing the exfoliated single-layers against reaggregation [124].

Unlike ultrasonication-assisted methods that rely primarily on mechanical forces, solvent-assisted exfoliation can proceed via a more passive mechanism under mild stirring or incubation, though in many cases it is combined with sonication to improve delamination efficiency. Solvents such as NMP [125], DMF [126], DMSO [127], and isopropanol [128] have been extensively employed due to their favorable surface energy parameters and ability to form stable colloidal dispersions of exfoliated MoS_2. The role of solvent polarity, boiling point, and viscosity also significantly influences the exfoliation outcome, with high-boiling, polar aprotic solvents demonstrating particular success in stabilizing monolayer and few-layer MoS_2 [127].

Despite its conceptual simplicity, solvent-assisted exfoliation is not without limitations. The process often yields low concentrations of exfoliated material and requires lengthy processing durations to achieve satisfactory delamination/exfoliation levels. Furthermore, the removal of residual solvent—particularly those with high boiling points—can be challenging and may compromise the structural and chemical integrity of the nanoparticles. Environmental and health concerns associated with certain solvents (e.g., NMP and DMF) have also prompted the search for greener alternatives, including water-based or bio-derived solvents, although these often suffer from reduced exfoliation efficiency [129].

Recent research has sought to improve the performance of solvent-assisted exfoliation through a range of innovations, including the use of solvent mixtures, phase-transfer agents, and post-exfoliation centrifugation techniques to enrich specific flake thickness distributions [130]. Additionally, there is growing interest in coupling solvent selection with computational modeling of solvent-material interactions to rationally design more effective exfoliation systems. Although challenges remain, solvent-assisted exfoliation continues to be a foundational technique for producing 2D MoS_2 with desirable morphological and physicochemical characteristics for applications in catalysis, photodetectors, transistors, and energy storage devices.

3 Modification Strategies of MoS_2 Nanoparticles

Owing to their structural versatility and tunable electronic properties, MoS_2 nanoparticles have emerged as highly promising materials for a broad spectrum of applications, most notably in catalysis, energy storage, optoelectronic devices, and environmental remediation. However, to fully exploit their potential, it is essential to modify and functionalize MoS_2 through various chemical and structural alterations [131]. These modification strategies are primarily focused on doping with metal and non-metallic elements, morphology control, and defect engineering, each of which are pivotal in enhancing the physicochemical characteristics of MoS_2 [132]. Modifications in MoS_2 are aimed at improving charge transport, optimizing catalytic activity, increasing mechanical stability, and expanding its environmental durability, thus making it more suitable for real-world applications [133].

3.1 Doping and Functionalization

Doping and functionalization strategies have been widely adopted to tailor the electronic, optical, and catalytic properties of MoS_2. By incorporating foreign atoms or molecular species into the MoS_2 lattice, these approaches effectively modulate its bandgap, enhance carrier mobility, and modify its chemical reactivity. Metal doping is particularly effective for enhancing catalytic activity and electrical conductivity, whereas non-metallic functionalization is primarily employed to regulate surface chemistry, hydrophilicity, and environmental stability.

3.1.1 Metal Doping

Incorporating transition metals into the MoS_2 lattice has effectively modified its intrinsic properties. The doping of MoS_2 with elements such as vanadium (V) [134, 135], titanium (Ti) [136, 137], nickel (Ni) [138–140], copper (Cu) [141, 142], palladium (Pd) [143, 144], and platinum (Pt) [145–147] significantly alters its electronic structure, surface chemistry, and catalytic performance. For instance, V- and Ti-doped MoS_2 significantly enhanced HER activity, as demonstrated by both density functional theory (DFT) calculations and experimental studies, attributed to an increased number of active sites for hydrogen adsorption [135, 136]. The introduction of transition metals leads to localized modifications in the electronic density of states, resulting in improved charge transfer kinetics and superior catalytic efficiency [133].

Nickel doping has been particularly advantageous for energy storage applications, as Ni-MoS_2 exhibit higher electrochemical stability [148] and increased electron mobility, making it a promising material as an anode for both lithium-ion and sodium-ion batteries [149]. Studies have shown that the incorporation of nickel into MoS_2 layers enhances its electrical conductivity by reducing charge recombination and facilitating ion diffusion, thereby improving cyclic performance and capacity retention in battery systems [150, 151]. Similarly, platinum and palladium doping have been explored to improve MoS_2's electrocatalytic activity for HER, where the strong metal-support interactions result in significant reductions in overpotential, thereby enhancing hydrogen production efficiency [146, 152].

Significant enhancements in the electrocatalytic activity of MoS_2 for key energy reactions have been theoretically predicted via transition metal doping. A DFT study by Joseph and Dabhi [153] identified Nb-doped MoS_2 as a highly efficient HER catalyst, with a calculated Gibbs free energy change (ΔG_H*) of only 0.06 eV. The same study demonstrated that Co-doping substantially improves performance for OER and ORR, yielding favorable overpotentials of 1.00 and 0.77 V, respectively.

Several synthesis techniques have been developed to introduce metal dopants into MoS_2 nanoparticles. Hydrothermal and solvothermal methods remain the most commonly used, allowing metal ions to be incorporated into the precursor solution during the crystal growth process [154]. CVD has also been widely applied, where metal-containing vapor-phase precursors are co-deposited onto substrates under controlled temperature and pressure conditions, leading to uniform doping [152]. Electrochemical deposition, in which dopant ions are introduced into MoS_2 films under applied voltage conditions, has further demonstrated its effectiveness in achieving precise and localized doping [155], particularly for applications in flexible electronic devices [156].

3.1.2 Non-metallic Functionalization

Beyond metal doping, the functionalization of MoS_2 with non-metallic species has emerged as a powerful strategy to fine-tune its electronic and surface properties [157, 158]. The incorporation of non-metallic dopant ions such as nitrogen (N) [158], phosphorus (P) [159–161], boron (B) [162], oxygen (O) [163, 164], leads to significant modifications in the band structure, surface reactivity, and chemical stability of MoS_2 [165]. Specifically, nitrogen doping has received attention for its effectiveness in boosting the catalytic performance of MoS_2 during electrocatalytic water splitting [166]. The substitution of sulfur atoms with nitrogen creates new electronic states near the Fermi level, thereby facilitating charge transfer and lowering HER energy barriers [167]. This leads to nitrogen-doped MoS_2 nanoparticles exhibiting a significant reduction in overpotential and improved long-term catalytic stability, as confirmed by experimental studies, making them ideal candidates for the sustainable production of hydrogen [168, 169].

Phosphorus doping has been extensively investigated for its impact on MoS_2's gas sensing capabilities by increasing surface-active sites, which in turn enhances the adsorption-desorption of gases (such as NH_3 and NO_2) [160, 161]. This leads to MoS_2-based gas sensors with better sensitivity, selectivity, and faster response times, making them crucial for environmental and industrial safety [160].

The non-metallic modification of MoS_2 also involves controlled defect engineering, particularly through the creation of sulfur vacancies. Removing sulfur atoms from the MoS_2 lattice generates reactive sites that enhance catalytic performance and adsorption properties. This strategy has been widely applied in environmental remediation, with sulfur-deficient MoS_2 materials showing superior heavy metal adsorption capabilities as well as enhanced photocatalytic degradation of organic pollutants [77, 170].

Another effective functionalization strategy involves modifying MoS_2 with organic molecules, including amines [171] and carboxylates [172]. This improves its solubility in various solvents and enhance its interfacial compatibility with polymers and composites. Hence, the organic functionalization has been particularly beneficial in the development of MoS_2-based nanocomposites for flexible electronics and biomedical applications [131, 157].

3.2 *Defect Engineering and Vacancy Creation*

Defect engineering is a powerful, well-established tool for tailoring the electronic, optical, and catalytic properties of MoS_2 nanostructures. Introducing vacancies, strain, and structural disorder into the MoS_2 lattice enhances the material's reactivity, charge transport characteristics, and overall functional performance. Among the different types of defects, sulfur vacancies have been extensively studied owing to their strong influence on catalytic activity and electronic conductivity [173].

3.2.1 Sulfur Vacancies

Sulfur vacancies are one of the most critical defects in MoS_2, as they create localized states in the electronic band structure that significantly enhance catalytic performance [174]. These vacancies introduce active sites that improve the efficiency of the HER [175] and CO_2 reduction reaction (CO_2RR) [176], making MoS_2 a promising candidate for sustainable energy applications [177].

Several methods have been employed to introduce sulfur vacancies into MoS_2. Thermal annealing in a reducing environment, such as hydrogen or argon, has been widely used to selectively remove sulfur atoms while maintaining the overall crystallinity of the material [178, 179]. Plasma treatment, particularly reactive ion etching, offers a more controlled approach by selectively etching sulfur atoms under precisely regulated plasma conditions [179]. Additionally, electrochemical reduction techniques have been explored, where an applied potential induces the formation of vacancies by altering the oxidation state of MoS_2 [175].

Sulfur-deficient MoS_2 has demonstrated superior catalytic performance. This enhanced reactivity stems from the increased number of exposed molybdenum sites and the favorable alteration of the electronic density of states induced by the vacancies. These modified sites and their altered electronic structure serve as highly effective active centers for redox reactions, optimizing both adsorption and charge transfer kinetics [180], which has been particularly beneficial in applications such as electrochemical water splitting, CO_2 conversion, and fuel cell technologies [181].

3.2.2 Edge Functionalization and Phase Transition Control

The edge sites of MoS_2 exhibit distinct chemical properties compared to basal planes, making them highly reactive for catalytic applications [11]. Functionalizing these edge sites with metal atoms, oxygen groups, or organic ligands has been shown to enhance their reactivity and broaden their functional scope. Metal-functionalized MoS_2 has demonstrated improved catalytic activity, particularly in HER and oxygen reduction reactions (ORR), where edge functionalization facilitates efficient charge transfer and reaction kinetics [182].

Another crucial aspect of defect engineering controls phase transitions in MoS_2 nanostructures, allowing the interconversion between its semiconducting 2H and metallic 1T phases [11]. Particularly, the 1T phase exhibits significantly higher electrical conductivity, making it an attractive option for applications in energy storage and electronic devices [183, 184]. Hence, this phase can be stabilized through lithium intercalation by reduction, allowing for tunable electronic properties [185]. Recent advancements in phase engineering have demonstrated that controlled phase transitions can be induced via external stimuli such as strain, pressure, or chemical treatments, further expanding the applicability of MoS_2-based materials in nanoelectronics and electrocatalysis [186, 187].

4 Conclusions and Outlook

In summary, the progress in the controlled synthesis and functionalization of MoS_2 nanostructures over the last decade has positioned them as leading materials for diverse advanced technologies. This chapter has highlighted the evolution of chemical and physical synthesis methodologies that enable precise control over morphology and phase and yield high-quality layered nanoparticles. Crucially, the strategic implementation of modification strategies, including metal doping to boost catalytic activity and conductivity, non-metallic functionalization to fine-tune electronic and surface properties, and defect engineering through sulfur vacancies and edge functionalization to enhance reactivity, has been pivotal in expanding the functional performance of MoS_2-based materials. These innovations have opened new avenues in nanoelectronics, electrocatalysis, energy storage, and biomedicine. The next frontier in MoS_2 research will undoubtedly involve the rational design and fabrication of hybrid and composite systems. Synergistically combining MoS_2 with other 2D materials, metals, or polymers to form heterostructures, thereby unlocking unprecedented functionalities and driving the development of next-generation advanced functional devices.

Acknowledgments The authors gratefully acknowledge the support from the Brazilian agencies: Fundação Araucária, CAPES (Finance code 001) and CNPq (FW—grant 303776/2023-4).

References

1. Kovalenko, M.V., Manna, L., Cabot, A., Hens, Z., Talapin, D.V., Kagan, C.R., Klimov, V.I., Rogach, A.L., Reiss, P., Milliron, D.J., Guyot-Sionnnest, P., Konstantatos, G., Parak, W.J., Hyeon, T., Korgel, B.A., Murray, C.B., Heiss, W.: Prospects of nanoscience with nanocrystals. ACS Nano. **9**, 1012–1057 (2015). https://doi.org/10.1021/NN506223H/ASSET/IMAGES/LARGE/NN-2014-06223H_0019.JPEG
2. Burda, C., Chen, X., Narayanan, R., El-Sayed, M.A.: Chemistry and properties of nanocrystals of different shapes. Chem. Rev. **105**, 1025–1102 (2005). https://doi.org/10.1021/cr030063a
3. Pinto, F.M., Dey, S., Taft, C.A., La Porta, F.A.: Synthesis of colloidal semiconductor nanocrystals for enhanced optoelectronic technologies. Eng. Mater. **Part F3949**, 1–47 (2024). https://doi.org/10.1007/978-3-031-75984-0_1
4. Splendiani, A., Sun, L., Zhang, Y., Li, T., Kim, J., Chim, C.Y., Galli, G., Wang, F.: Emerging photoluminescence in monolayer MoS2. Nano Lett. **10**, 1271–1275 (2010). https://doi.org/10.1021/NL903868W/ASSET/IMAGES/LARGE/NL-2009-03868W_0004.JPEG
5. Pinto, F.M., de Conti, M.C.M.D., Pereira, W.S., Sczancoski, J.C., Medina, M., Corradini, P.G., de Brito, J.F., Nogueira, A.E., Góes, M.S., Ferreira, O.P., Mascaro, L.H., Wypych, F., La Porta, F.A.: Recent advances in layered MX2-based materials (M = Mo, W and X = S, Se, Te) for emerging optoelectronic and photo(electro)catalytic applications. Catalysts. **14**, 388 (2024). https://doi.org/10.3390/CATAL14060388
6. Mak, K.F., Lee, C., Hone, J., Shan, J., Heinz, T.F.: Atomically thin MoS2: a new direct-gap semiconductor. Phys. Rev. Lett. **105**, 136805 (2010). https://doi.org/10.1103/PHYSREVLETT.105.136805/FIGURES/4/THUMBNAIL
7. Chhowalla, M., Shin, H.S., Eda, G., Li, L.J., Loh, K.P., Zhang, H.: The chemistry of two-dimensional layered transition metal dichalcogenide nanoparticles. Nat. Chem. **5**(4), 263–275 (2013). https://doi.org/10.1038/nchem.1589
8. Wang, Q.H., Kalantar-Zadeh, K., Kis, A., Coleman, J.N., Strano, M.S.: Electronics and optoelectronics of two-dimensional transition metal dichalcogenides. Nat. Nanotechnol. **7**(11), 699–712 (2012). https://doi.org/10.1038/nnano.2012.193
9. Das, S., Swain, G., Parida, K.: One step towards the 1T/2H-MoS2 mixed phase: a journey from synthesis to application. Mater. Chem. Front. **5**, 2143–2172 (2021). https://doi.org/10.1039/D0QM00802H
10. Huang, Q., Shen, J., Lu, Y., Ye, R., Gong, S.: Insights into the structural evolution of MoS2 from the semiconductive 2H to metallic 1T phase. J. Phys. Chem. C. **127**, 17406–17414 (2023). https://doi.org/10.1021/ACS.JPCC.3C03254/SUPPL_FILE/JP3C03254_SI_001.PDF
11. Pinto, F.M., La Porta, F.A., Wypych, F.: 1T-MoS_2: a brief review of the properties and applications of this metallic polymorph of the molybdenum disulfide. Quim Nova. **48**, e-20250085 (2025). https://doi.org/10.21577/0100-4042.20250085

12. Luxa, J., Spejchalová, L., Jakubec, I., Sofer, Z.: MoS2 stacking matters: 3R polytype significantly outperforms 2H MoS2 for the hydrogen evolution reaction. Nanoscale. **13**, 19391–19398 (2021). https://doi.org/10.1039/D1NR03284D
13. Toh, R.J., Sofer, Z., Luxa, J., Sedmidubský, D., Pumera, M.: 3R phase of MoS2 and WS2 outperforms the corresponding 2H phase for hydrogen evolution. Chem. Commun. **53**, 3054–3057 (2017). https://doi.org/10.1039/C6CC09952A
14. Ren, T., Tian, W., Shen, Q., Yuan, Z., Chen, D., Li, N., Lu, J.: Enhanced piezocatalysis of polymorphic few-layered MoS2 nanoparticles by phase engineering. Nano Energy. **90**, 106527 (2021). https://doi.org/10.1016/J.NANOEN.2021.106527
15. Saber, M.R., Khabiri, G., Maarouf, A.A., Ulbricht, M., Khalil, A.S.G.: A comparative study on the photocatalytic degradation of organic dyes using hybridized 1T/2H, 1T/3R and 2H MoS2 nano-sheets. RSC Adv. **8**, 26364–26370 (2018). https://doi.org/10.1039/C8RA05387A
16. Coleman, J.N., Lotya, M., O'Neill, A., Bergin, S.D., King, P.J., Khan, U., Young, K., Gaucher, A., De, S., Smith, R.J., Shvets, I.V., Arora, S.K., Stanton, G., Kim, H.Y., Lee, K., Kim, G.T., Duesberg, G.S., Hallam, T., Boland, J.J., Wang, J.J., Donegan, J.F., Grunlan, J.C., Moriarty, G., Shmeliov, A., Nicholls, R.J., Perkins, J.M., Grieveson, E.M., Theuwissen, K., McComb, D.W., Nellist, P.D., Nicolosi, V.: Two-dimensional nanoparticles produced by liquid exfoliation of layered materials. Science. **331**(2011), 568–571 (1979). https://doi.org/10.1126/SCIENCE.1194975/SUPPL_FILE/COLEMAN.SOM.PDF
17. Zhan, Y., Liu, Z., Najmaei, S., Ajayan, P.M., Lou, J.: Large-area vapor-phase growth and characterization of MoS2 atomic layers on a SiO2 substrate. Small. **8**, 966–971 (2012). https://doi.org/10.1002/SMLL.201102654
18. Thomas, N., Mathew, S., Nair, K.M., O'Dowd, K., Forouzandeh, P., Goswami, A., McGranaghan, G., Pillai, S.C.: 2D MoS2: structure, mechanisms, and photocatalytic applications. Mater. Today Sustain. **13**, 100073 (2021). https://doi.org/10.1016/J.MTSUST.2021.100073
19. Burse, S.R., Tyagaraj, H.B., Safarkhani, M., Marje, S.J., Gagankumar, S.K., Al Ghaferi, A., Alhajri, E., Chodankar, N.R., Huh, Y.S., Han, Y.K.: Unleashing potential: engineering advancements in two-dimensional MoS2 for improved energy applications. Adv. Compos. Hybrid Mater. **8**, 1–54 (2025). https://doi.org/10.1007/S42114-025-01289-Y
20. Tian, Y., Zhao, X., Shen, L., Meng, F., Tang, L., Deng, Y., Wang, Z.: Synthesis of amorphous MoS2 nanospheres by hydrothermal reaction. Mater. Lett. **60**, 527–529 (2006). https://doi.org/10.1016/J.MATLET.2005.09.029
21. Wang, L., Ma, Y., Yang, M., Qi, Y.: Hierarchical hollow MoS2 nanospheres with enhanced electrochemical properties used as an electrode in supercapacitor. Electrochim. Acta. **186**, 391–396 (2015). https://doi.org/10.1016/J.ELECTACTA.2015.10.130
22. Park, S.K., Yu, S.H., Woo, S., Ha, J., Shin, J., Sung, Y.E., Piao, Y.: A facile and green strategy for the synthesis of MoS2 nanospheres with excellent Li-ion storage properties. CrystEngComm. **14**, 8323–8325 (2012). https://doi.org/10.1039/C2CE26447A
23. Ye, Z., Yang, J., Li, B., Shi, L., Ji, H., Song, L., Xu, H.: Amorphous molybdenum sulfide/carbon nanotubes hybrid nanospheres prepared by ultrasonic spray pyrolysis for electrocatalytic hydrogen evolution. Small. **13**, 1700111 (2017). https://doi.org/10.1002/SMLL.201700111; WGROUP:STRING:PUBLICATION
24. Choi, S.H., Ko, Y.N., Lee, J.K., Kang, Y.C.: 3D MoS2–graphene microspheres consisting of multiple nanospheres with superior sodium ion storage properties. Adv. Funct. Mater. **25**, 1780–1788 (2015). https://doi.org/10.1002/ADFM.201402428
25. Waskito, I.S., Kurniawan, B., Amal, M.I., Hanifuddin, M.: The effect of precursors concentration on the structural properties of MoS2 nanosheet-microsphere synthesized via hydrothermal route. IOP Conf. Ser. Mater. Sci. Eng. **546**, 042048 (2019). https://doi.org/10.1088/1757-899X/546/4/042048
26. Lee, C.M., Park, G.C., Ahn, C.H., Choi, J.W., Park, C.H., Cho, H.K., Lim, J.H., Joo, J.: Effects of precursor concentration on dimensional size, defect state, and gas sensing perfor-

mance of MoS2 sheets synthesized by hydrothermal method. Phys. Status Solidi (A) Appl. Mater. Sci. **215**, 1800079 (2018). https://doi.org/10.1002/PSSA.201800079
27. Remskar, M., Mrzel, A., Skraba, Z., Jesih, A., Ceh, M., Demšar, J., Stadelmann, P., Lévy, F., Mihailovic, D.: Self-assembly of subnanometer-diameter single-wall MoS2 nanotubes. Science. **292**(2001), 479–481 (1979). https://doi.org/10.1126/SCIENCE.1059011;REQUESTEDJOURNAL:JOURNAL:SCIENCE;CTYPE:STRING:JOURNAL
28. Chen, J., Kuriyama, N., Yuan, H., Takeshita, H.T., Sakai, T.: Electrochemical hydrogen storage in MoS2 nanotubes. J. Am. Chem. Soc. **123**, 11813–11814 (2001). https://doi.org/10.1021/JA017121Z/ASSET/IMAGES/LARGE/JA017121ZH00001.JPEG
29. Okamoto, T., Shimizu, T., Takase, K., Ito, T., Shingubara, S.: Formation of MoS2 nanostructure arrays using anodic aluminum oxide template. Micro Nano Eng. **9**, 100071 (2020). https://doi.org/10.1016/J.MNE.2020.100071
30. Jia, Y., Yin, G., Lin, Y., Ma, Y.: Recent progress on hierarchical MoS2 nanostructures for electrochemical energy storage. CrystEngComm. **24**, 2314–2326 (2022). https://doi.org/10.1039/D1CE01439K
31. Tang, Q., Jiang, D.E.: Stabilization and band-gap tuning of the 1T-MoS2 monolayer by covalent functionalization. Chem. Mater. **27**, 3743–3748 (2015). https://doi.org/10.1021/ACS.CHEMMATER.5B00986/ASSET/IMAGES/LARGE/CM-2015-00986X_0007.JPEG
32. Sen, P.: Computational screening of layered metal chalcogenide materials for HER electrocatalysts, and its synergy with experiments. J. Phys. Condens. Matter. **36**, 223002 (2024). https://doi.org/10.1088/1361-648X/AD2D45
33. Voiry, D., Yang, J., Chhowalla, M.: Recent strategies for improving the catalytic activity of 2D TMD nanoparticles toward the hydrogen evolution reaction. Adv. Mater. **28**, 6197–6206 (2016). https://doi.org/10.1002/adma.201505597
34. Voiry, D., Yamaguchi, H., Li, J., Silva, R., Alves, D.C.B., Fujita, T., Chen, M., Asefa, T., Shenoy, V.B., Eda, G., Chhowalla, M.: Enhanced catalytic activity in strained chemically exfoliated WS2 nanoparticles for hydrogen evolution. Nat. Mater. **12**(9), 850–855 (2013). https://doi.org/10.1038/nmat3700
35. Tongay, S., Zhou, J., Ataca, C., Liu, J., Kang, J.S., Matthews, T.S., You, L., Li, J., Grossman, J.C., Wu, J.: Broad-range modulation of light emission in two-dimensional semiconductors by molecular physisorption gating. Nano Lett. **13**, 2831–2836 (2013). https://doi.org/10.1021/NL4011172/SUPPL_FILE/NL4011172_SI_001.PDF
36. Kamat, P.V.: Graphene-based nanoarchitectures. Anchoring semiconductor and metal nanoparticles on a two-dimensional carbon support. J. Phys. Chem. Lett. **1**, 520–527 (2010). https://doi.org/10.1021/JZ900265J/ASSET/IMAGES/LARGE/JZ-2009-00265J_0012.JPEG
37. Late, D.J., Liu, B., Matte, H.S.S.R., Dravid, V.P., Rao, C.N.R.: Hysteresis in single-layer MoS2 field effect transistors. ACS Nano. **6**, 5635–5641 (2012). https://doi.org/10.1021/NN301572C/SUPPL_FILE/NN301572C_SI_001.PDF
38. Eda, G., Maier, S.A.: Two-dimensional crystals: managing light for optoelectronics. ACS Nano. **7**, 5660–5665 (2013). https://doi.org/10.1021/NN403159Y/ASSET/IMAGES/LARGE/NN-2013-03159Y_0002.JPEG
39. Tan, C., Zhang, H.: Wet-chemical synthesis and applications of non-layer structured two-dimensional nanomaterials. Nat. Commun. **6**(1), 1–13 (2015). https://doi.org/10.1038/ncomms8873
40. Huo, J., Ge, R., Liu, Y., Guo, J., Lu, L., Chen, W., Liu, C., Gao, H., Liu, H.: Recent advances of two–dimensional molybdenum disulfide based materials: synthesis, modification and applications in energy conversion and storage. Sustain. Mater. Technol. **24**, e00161 (2020). https://doi.org/10.1016/J.SUSMAT.2020.E00161
41. Mohan, B., Boruah, U., Sonkar, R., Mondal, N.J., Chowdhury, D.: 2D-MoX2 (X = S, Se, Te) and their nanocomposite toward sensing application: a review. Part. Part. Syst. Charact. **42**, 2400156 (2025). https://doi.org/10.1002/PPSC.202400156

42. Sun, J., Li, X., Guo, W., Zhao, M., Fan, X., Dong, Y., Xu, C., Deng, J., Fu, Y.: Synthesis methods of two-dimensional MoS2: a brief review. Crystals. **7**, 198 (2017). https://doi.org/10.3390/CRYST7070198
43. de Conti, M.C.M.D., de Castro, A.A., Assis, L.C., Lima, N.M., Escriba, A., Nepovimova, E., Kuca, K., Ramalho, T.C., La Porta, F.A.: Carbon dots in the center of the spotlight: a full evaluation of their synthesis and understanding of their fundamental properties and applications. Mater. Today Sustain. **27**, 100937 (2024). https://doi.org/10.1016/J.MTSUST.2024.100937
44. Nakata, M.M., Mazzo, T.M., Casali, G.P., La Porta, F.A., Longo, E.: A large red-shift in the photoluminescence emission of Mg1−xSrxTiO3. Chem. Phys. Lett. **622**, 9–14 (2015). https://doi.org/10.1016/J.CPLETT.2015.01.011
45. Raza, A., Hassan, J.Z., Ikram, M., Ali, S., Farooq, U., Khan, Q., Maqbool, M.: Advances in liquid-phase and intercalation exfoliations of transition metal dichalcogenides to produce 2D framework. Adv. Mater. Interfaces. **8**, 2002205 (2021). https://doi.org/10.1002/ADMI.202002205
46. de Conti, M.C.M.D., Dey, S., Pottker, W.E., La Porta, F.A.: An overview into advantages and applications of conventional and unconventional hydro(solvo)thermal approaches for novel advanced materials design. Mater. Today Sustain. **23**, 100458 (2023). https://doi.org/10.1016/J.MTSUST.2023.100458
47. Huang, H., Liu, N., Wang, X., Zhong, M., Huang, X.: Application of hydrothermal and solvothermal method in synthesis of MoS2. Rev. Chim. **73**, 26–35 (2022). https://doi.org/10.37358/RC.22.4.8544
48. Luo, L., Shi, M., Zhao, S., Tan, W., Lin, X., Wang, H., Jiang, F. Hydrothermal synthesis of MoS2 with controllable morphologies and its adsorption properties for bisphenol A. J. Saudi Chem. Soc. **23**, 762–773 (2019). https://doi.org/10.1016/J.JSCS.2019.01.005
49. Yang, X., Wen, L., Huang, H., Wang, Y., Wei, L., Yang, J.: Hydrothermal synthesis of MoS2 nanoflowers and its rapid adsorption of tetracycline. Solid State Sci. **133**, 107014 (2022). https://doi.org/10.1016/J.SOLIDSTATESCIENCES.2022.107014
50. Nayak, D., Kumar, A., Thangavel, R.: Synthesis of MoS2 nanoflowers for photocatalytic degradation of organic dyes. ACS Appl. Nano Mater. **6**, 19476–19490 (2023). https://doi.org/10.1021/ACSANM.3C04316/SUPPL_FILE/AN3C04316_SI_001.PDF
51. Liu, C., Yang, P., Zhang, X.: MoS2 nanoparticles grown homogeneously on hollow carbon spheres for efficient hydrogen evolution reaction and supercapacitor. Int. J. Hydrog. Energy. **97**, 1348–1356 (2025). https://doi.org/10.1016/J.IJHYDENE.2024.12.008
52. Gu, W., Yan, Y., Zhang, C., Ding, C., Xian, Y.: One-step synthesis of water-soluble MoS2 quantum dots via a hydrothermal method as a fluorescent probe for hyaluronidase detection. ACS Appl. Mater. Interfaces. **8**, 11272–11279 (2016). https://doi.org/10.1021/ACSAMI.6B01166/SUPPL_FILE/AM6B01166_SI_001.PDF
53. Huang, H.E., Liu, N.A., Wang, X., Zhong, M., Huang, X.: Application of hydrothermal and solvothermal method in synthesis of MoS2 (n.d.). https://doi.org/10.37358/Mat.Plast.1964
54. Simonenko, T.L., Simonenko, N.P., Zemlyanukhin, A.A., Gorobtsov, P.Y., Simonenko, E.P.: Hydrothermal synthesis of hierarchically organized MoS2 and formation of films based on it. Russ. J. Inorg. Chem. **69**(14), 1–12 (2024). https://doi.org/10.1134/S0036023624603039/FIGURES/9
55. Zhang, M., Wang, S., Li, Z., Liu, C., Miao, R., He, G., Zhao, M., Xue, J., Xia, Z., Wang, Y., Sun, Z., Lv, J.: Hydrothermal synthesis of MoS2 nanosheet loaded TiO2 nanoarrays for enhanced visible light photocatalytic applications. RSC Adv. **9**, 3479–3485 (2019). https://doi.org/10.1039/C8RA09348B
56. Strachan, J., Masters, A.F., Maschmeyer, T.: Critical review: hydrothermal synthesis of 1T-MoS2—an important route to a promising material. J. Mater. Chem. A Mater. **9**, 9451–9461 (2021). https://doi.org/10.1039/D1TA01230D
57. Xu, M.Z., Li, Q., Lv, Y.Y., Yuan, Z.M., Guo, Y.X., Jiang, H.J., Gao, J.W., Di, J., Song, P., Kang, L.X., Zheng, L., Zhang, Z.Y., Zhao, W., Wang, X.W., Liu, Z.: Surfactant-assisted hydrothermal synthesis of MoS2 micro-pompon structure with enhanced photocatalytic

performance under visible light. Tungsten. **2**, 203–213 (2020). https://doi.org/10.1007/S42864-020-00050-W/METRICS
58. Li, M., Wang, D., Li, J., Pan, Z., Ma, H., Jiang, Y., Tian, Z., Lu, A.: Surfactant-assisted hydrothermally synthesized MoS2 samples with controllable morphologies and structures for anthracene hydrogenation. Chin. J. Catal. **38**, 597–606 (2017). https://doi.org/10.1016/S1872-2067(17)62779-7
59. Liu, W., He, S., Wang, Y., Dou, Y., Pan, D., Feng, Y., Qian, G., Xu, J., Miao, S.: PEG-assisted synthesis of homogeneous carbon nanotubes-MoS2-carbon as a counter electrode for dye-sensitized solar cells. Electrochim. Acta. **144**, 119–126 (2014). https://doi.org/10.1016/J.ELECTACTA.2014.08.075
60. de Oliveira, M.L., Rodrigues, L.M., Silva da Veiga, M.A.M., Souza, L.R.R.: Advancing nanomaterial synthesis: harnessing green chemistry for sustainable innovation. Green Anal. Chem. **11**, 100148 (2024). https://doi.org/10.1016/J.GREEAC.2024.100148
61. Wang, S., Ding, J., Wang, C., Li, W., Chen, Z., Liu, C., Chen, F.: Construction of molybdenum disulfide/biological structure carbon composite photocatalysts and their photocatalytic hydrogen production. Front. Mater. **9**, 889499 (2022). https://doi.org/10.3389/FMATS.2022.889499/BIBTEX
62. Lu, X.W., Li, Z., Yang, C.K., Mou, W., Jiao, L.: Synthesis of uniform two-dimensional MoS2 films via thermal evaporation. Nano Res. **17**, 3217–3223 (2024). https://doi.org/10.1007/S12274-023-6114-Z/METRICS
63. Ye, Z., Tan, C., Huang, X., Ouyang, Y., Yang, L., Wang, Z., Dong, M.: Emerging MoS2 wafer-scale technique for integrated circuits. Nano-Micro Lett. **15**(1), 1–42 (2023). https://doi.org/10.1007/S40820-022-01010-4
64. Koroteev, V.O., Stolyarova, S.G., Kotsun, A.A., Modin, E., Makarova, A.A., Shubin, Y.V., Plyusnin, P.E., Okotrub, A.V., Bulusheva, L.G.: Nanoscale coupling of MoS2 and graphene via rapid thermal decomposition of ammonium tetrathiomolybdate and graphite oxide for boosting capacity of Li-ion batteries. Carbon N Y. **173**, 194–204 (2021). https://doi.org/10.1016/J.CARBON.2020.10.097
65. Kovács, T.N., Hunyadi, D., de Lucena, A.L.A., Szilágyi, I.M.: Thermal decomposition of ammonium molybdates. J. Therm. Anal. Calorim. **124**, 1013–1021 (2016). https://doi.org/10.1007/S10973-015-5201-0/FIGURES/14
66. Yoo, K., Koo, W.B., Kim, H., Lee, S.H.: The self-reduction during the thermal decomposition of an ammonium molybdate. Minerals. **13**, 133 (2023). https://doi.org/10.3390/MIN13020133
67. Vinoba, M., Navvamani, R., Al-Sheeha, H.: Controllable thermal conversion of thiomolybdate to active few-layer MoS2 on alumina for efficient hydrodesulfurization. SN Appl. Sci. **1**, 1–11 (2019). https://doi.org/10.1007/S42452-019-0361-3/FIGURES/10
68. Wypych, F.: Dissulfeto de molibdênio, um material multifuncional e surpreendente. Quim Nova. **25**, 83–88 (2002). https://doi.org/10.1590/S0100-40422002000100015
69. Wypych, F., Schoellhorn, R.: ChemInform abstract: 1T-MoS2, a new metallic modification of molybdenum disulfide. ChemInform. **24** (1993). https://doi.org/10.1002/CHIN.199303033
70. Chen, R., Bresser, D., Saraf, M., Gerlach, P., Balducci, A., Kunz, S., Schröder, D., Passerini, S., Chen, J.: A comparative review of electrolytes for organic-material-based energy-storage devices employing solid electrodes and redox fluids. ChemSusChem. **13**, 2205–2219 (2020). https://doi.org/10.1002/CSSC.201903382
71. Pameté, E., Köps, L., Kreth, F.A., Pohlmann, S., Varzi, A., Brousse, T., Balducci, A., Presser, V.: The many deaths of supercapacitors: degradation, aging, and performance fading. Adv. Energy Mater. **13**, 2301008 (2023). https://doi.org/10.1002/AENM.202301008;PAGE:STRING:ARTICLE/CHAPTER
72. Besmann, T.M., Stinton, D.P., Lowden, R.A.: Chemical vapor deposition techniques. MRS Bull. **13**, 45–51 (1988). https://doi.org/10.1557/S0883769400063910/METRICS

73. Lei, M., Wang, P., Ke, X., Xie, J., Yue, M., Zhao, M., Zhang, K., Dong, Y., Xu, Q., Zou, C., Wang, S., Zhang, L.: Low-temperature chemical vapor deposition growth of 2D materials. Des. Electron. **3**, e43 (2024). https://doi.org/10.1002/ELT2.43;CTYPE:STRING:JOURNAL
74. Zhang, Q., Geng, D., Hu, W.: Chemical vapor deposition for few-layer two-dimensional materials. SmartMat. **4**, e1177 (2023). https://doi.org/10.1002/SMM2.1177
75. Guan, R., Duan, J., Yuan, A., Wang, Z., Yang, S., Han, L., Zhang, B., Li, D., Luo, B.: Chemical vapor deposition of clean and pure MoS2 crystals by the inhibition of MoO3−x intermediates. CrystEngComm. **23**, 146–152 (2021). https://doi.org/10.1039/D0CE01354D
76. Gao, Q., Lu, J., Chen, S., Chen, L., Xu, Z., Lin, D., Xu, S., Liu, P., Zhang, X., Cai, W., Zhang, C.: Chemical vapor deposition of uniform and large-domain molybdenum disulfide crystals on glass/Al2O3 substrates. Nanomaterials. **12**, 2719 (2022). https://doi.org/10.3390/NANO12152719/S1
77. Wree, J.L., Ciftyurek, E., Zanders, D., Boysen, N., Kostka, A., Rogalla, D., Kasischke, M., Ostendorf, A., Schierbaum, K., Devi, A.: A new metalorganic chemical vapor deposition process for MoS2 with a 1,4-diazabutadienyl stabilized molybdenum precursor and elemental sulfur. Dalton Trans. **49**, 13462–13474 (2020). https://doi.org/10.1039/D0DT02471F
78. Grabicki, N., Dumele, O., Sai, H., Powers-Riggs, N.E., Phelan, B.T., Sangji, M.H., Chapman, C.T., Passarelli, J.V., Dannenhoffer, A.J., Wasielewski, M.R., Stupp, S.I.: Polymorphism and optoelectronic properties in crystalline supramolecular polymers. Chem. Mater. **33**, 706–718 (2021). https://doi.org/10.1021/ACS.CHEMMATER.0C04123/SUPPL_FILE/CM0C04123_SI_003.CIF
79. Dumcenco, D., Ovchinnikov, D., Marinov, K., Lazić, P., Gibertini, M., Marzari, N., Sanchez, O.L., Kung, Y.C., Krasnozhon, D., Chen, M.W., Bertolazzi, S., Gillet, P., Morral, A.F.I., Radenovic, A., Kis, A.: Large-area epitaxial monolayer moS2. ACS Nano. **9**, 4611–4620 (2015). https://doi.org/10.1021/ACSNANO.5B01281/SUPPL_FILE/NN5B01281_SI_001.PDF
80. Jasinski, J.M., Meyerson, B.S., Scott, B.A.: Mechanistic studies of chemical vapor deposition. Annu. Rev. Phys. Chem. **38**, 109–140 (1987). https://doi.org/10.1146/ANNUREV.PC.38.100187.000545
81. Liu, N., Kim, J., Oh, J., Nguyen, Q.T., Sahu, B.B., Han, J.G., Kim, S.: Growth of multi-orientated polycrystalline MoS2 using plasma-enhanced chemical vapor deposition for efficient hydrogen evolution reactions. Nanomaterials. **10**, 1465 (2020). https://doi.org/10.3390/NANO10081465
82. Yi, K., Liu, D., Chen, X., Yang, J., Wei, D., Liu, Y., Wei, D.: Plasma-enhanced chemical vapor deposition of two-dimensional materials for applications. Acc. Chem. Res. **54**, 1011–1022 (2021). https://doi.org/10.1021/ACS.ACCOUNTS.0C00757/ASSET/IMAGES/LARGE/AR0C00757_0011.JPEG
83. Kim, T., Mun, J., Park, H., Joung, D., Diware, M., Won, C., Park, J., Jeong, S.H., Kang, S.W.: Wafer-scale production of highly uniform two-dimensional MoS2 by metal-organic chemical vapor deposition. Nanotechnology. **28**, 18LT01 (2017). https://doi.org/10.1088/1361-6528/AA6958
84. Cwik, S., Mitoraj, D., Mendoza Reyes, O., Rogalla, D., Peeters, D., Kim, J., Maria Schütz, H., Bock, C., Beranek, R., Devi, A., Cwik, S., Peeters, D., Kim, J., Devi, A., Mitoraj, D., Mendoza Reyes, O., Schütz, H.M., Beranek, R., Rogalla, D., Bock, C.: Direct growth of MoS2 and WS2 layers by metal organic chemical vapor deposition. Adv. Mater. Interfaces. **5**, 1800140 (2018). https://doi.org/10.1002/ADMI.201800140
85. Yi, K., Liu, D., Chen, X., Yang, J., Wei, D., Liu, Y., Wei, D.: Plasma-enhanced chemical vapor deposition of two-dimensional materials for applications. Acc. Chem. Res. **54**, 1011–1022 (2021). https://doi.org/10.1021/ACS.ACCOUNTS.0C00757/ASSET/IMAGES/MEDIUM/AR0C00757_0012.GIF
86. Ahn, C., Lee, J., Kim, H.U., Bark, H., Jeon, M., Ryu, G.H., Lee, Z., Yeom, G.Y., Kim, K., Jung, J., Kim, Y., Lee, C., Kim, T.: Low-temperature synthesis of large-scale molybdenum

disulfide thin films directly on a plastic substrate using plasma-enhanced chemical vapor deposition. Adv. Mater. **27**, 5223–5229 (2015). https://doi.org/10.1002/ADMA.201501678
87. Bo, Z., Yang, Y., Chen, J., Yu, K., Yan, J., Cen, K.: Plasma-enhanced chemical vapor deposition synthesis of vertically oriented graphene nanoparticles. Nanoscale. **5**, 5180–5204 (2013). https://doi.org/10.1039/C3NR33449J
88. Mawlong, L.P.L., Hoang, A.T., Chintalapalli, J., Ji, S., Lee, K., Kim, K., Ahn, J.H.: Reduced defect density in MOCVD-grown MoS2 by manipulating the precursor phase. ACS Appl. Mater. Interfaces. **15**, 47359–47367 (2023). https://doi.org/10.1021/ACSAMI.3C09027/SUPPL_FILE/AM3C09027_SI_001.PDF
89. Liu, L., Qiu, H., Wang, J., Xu, G., Jiao, L.: Atomic MoS2 monolayers synthesized from a metal–organic complex by chemical vapor deposition. Nanoscale. **8**, 4486–4490 (2016). https://doi.org/10.1039/C5NR09089J
90. Zhang, Z., Hoang, L., Hocking, M., Peng, Z., Hu, J., Zaborski, G., Reddy, P.D., Dollard, J., Goldhaber-Gordon, D., Heinz, T.F., Pop, E., Mannix, A.J.: Chemically tailored growth of 2D semiconductors via hybrid metal-organic chemical vapor deposition. ACS Nano. **18**(37), 25414–25424 (2024). https://doi.org/10.1021/ACSNANO.4C02164/SUPPL_FILE/NN4C02164_SI_001.PDF
91. Romanov, R.I., Zabrosaev, I.V., Chouprik, A.A., Yakubovsky, D.I., Tatmyshevskiy, M.K., Volkov, V.S., Markeev, A.M.: Temperature-dependent structural and electrical properties of metal-organic CVD MoS2 films. Nanomaterials. **13**, 2712 (2023). https://doi.org/10.3390/NANO13192712/S1
92. Eichfeld, S.M., Hossain, L., Lin, Y.C., Piasecki, A.F., Kupp, B., Birdwell, A.G., Burke, R.A., Lu, N., Peng, X., Li, J., Azcatl, A., McDonnell, S., Wallace, R.M., Kim, M.J., Mayer, T.S., Redwing, J.M., Robinson, J.A.: Highly scalable, atomically thin WSe2 grown via metal-organic chemical vapor deposition. ACS Nano. **9**, 2080–2087 (2015). https://doi.org/10.1021/NN5073286/SUPPL_FILE/NN5073286_SI_001.PDF
93. Chen, Y., Fang, F., Zhang, N.: Advance in additive manufacturing of 2D materials at the atomic and close-to-atomic scale. npj 2D Mater. Appl. **8**, 1–23 (2024). https://doi.org/10.1038/s41699-024-00456-x
94. Wu, R., Zhang, H., Ma, H., Zhao, B., Li, W., Chen, Y., Liu, J., Liang, J., Qin, Q., Qi, W., Chen, L., Li, J., Li, B., Duan, X.: Synthesis, modulation, and application of two-dimensional TMD heterostructures. Chem. Rev. **124**, 10112–10191 (2024). https://doi.org/10.1021/ACS.CHEMREV.4C00174/ASSET/IMAGES/LARGE/CR4C00174_0042.JPEG
95. Cantarella, M., Gorrasi, G., Di Mauro, A., Scuderi, M., Nicotra, G., Fiorenza, R., Scirè, S., Scalisi, M.E., Brundo, M.V., Privitera, V., Impellizzeri, G.: Mechanical milling: a sustainable route to induce structural transformations in MoS2 for applications in the treatment of contaminated water. Sci. Rep. **9**, 1–14 (2019). https://doi.org/10.1038/S41598-018-37798-8;TECHMETA=128,132,140,146;SUBJMETA=1018,301,357,551,639,925;KWRD=SYNTHESIS+AND+PROCESSING,TWO-DIMENSIONAL+MATERIALS
96. Ismail, K.B.M., Arun Kumar, M., Mahalingam, S., Kim, J., Atchudan, R.: Recent advances in molybdenum disulfide and its nanocomposites for energy applications: challenges and development. Materials. **16**, 4471 (2023). https://doi.org/10.3390/MA16124471
97. Saliba, M., Atanas, J.P., Howayek, T.M., Habchi, R.: Molybdenum disulfide, exfoliation methods and applications to photocatalysis: a review. Nanoscale Adv. **5**, 6787–6803 (2023). https://doi.org/10.1039/D3NA00741C
98. Ali, L., Subhan, F., Ayaz, M., ul Hassan, S.S., Byeon, C.C., Kim, J.S., Bungau, S.: Exfoliation of MoS2 quantum dots: recent progress and challenges. Nanomaterials. **12**, 3465 (2022). https://doi.org/10.3390/NANO12193465
99. Kumar, A., Rawat, R.K., Tripathi, S., Singh, S.P., Tripathi, D., Tripathi, S., Chauhan, P.: Captivating 2H-MoS2 nanoflowers for efficient NH3 detection and photocatalytic dye degradation. Colloids Surf. A Physicochem. Eng. Asp. **704**, 135414 (2025). https://doi.org/10.1016/J.COLSURFA.2024.135414

100. Py, M.A., Haering, R.R.: Structural destabilization induced by lithium intercalation in MoS2 and related compounds. Can. J. Phys. **61**, 76–84 (1983). https://doi.org/10.1139/P83-013;WEBSITE:WEBSITE:NRC-SITE;JOURNAL:JOURNAL:CJP;WGROUP:STRING:CSP
101. Gee, M.A., Frindt, R.F., Joensen, P., Morrison, S.R.: Inclusion compounds of MoS2. Mater. Res. Bull. **21**, 543–549 (1986). https://doi.org/10.1016/0025-5408(86)90108-X
102. Mulhern, P.J.: Lithium intercalation in crystalline LixMoS2. Can. J. Phys. **67**, 1049–1052 (1989). https://doi.org/10.1139/P89-184
103. Das, S., Tama, A.M., Dutta, S., Ali, M.S., Basith, M.A.: Facile high-yield synthesis of MoS2 nanoparticles with enhanced photocatalytic performance using ultrasound driven exfoliation technique. Mater. Res. Express. **6**, 125079 (2019). https://doi.org/10.1088/2053-1591/AB57DD
104. Galloni, M.G., Fabbrizio, V., Giannantonio, R., Falletta, E., Bianchi, C.L.: Applications and applicability of the cavitation technology. Curr. Opin. Chem. Eng. **48**, 101129 (2025). https://doi.org/10.1016/J.COCHE.2025.101129
105. Shchukin, D.G., Skorb, E., Belova, V., Möhwald, H.: Ultrasonic cavitation at solid surfaces. Adv. Mater. **23**, 1922–1934 (2011). https://doi.org/10.1002/ADMA.201004494;REQUESTEDJOURNAL:JOURNAL:15214095;WGROUP:STRING:PUBLICATION
106. Mukhopadhyay, T.K., Datta, A.: Disentangling the liquid phase exfoliation of two-dimensional materials: an "in silico" perspective. Phys. Chem. Chem. Phys. **22**, 22157–22179 (2020). https://doi.org/10.1039/D0CP03128C
107. Cai, M., Wei, Y., Li, Y., Li, X., Wang, S., Shao, G., Zhang, P.: 2D semiconductor nanoparticles for solar photocatalysis. EcoEnergy. **1**, 248–295 (2023). https://doi.org/10.1002/ECE2.16
108. Lovell, P.A., Schork, F.J.: Fundamentals of emulsion polymerization. Biomacromolecules. **21**, 4396–4441 (2020). https://doi.org/10.1021/ACS.BIOMAC.0C00769/SUPPL_FILE/BM0C00769_LIVESLIDES.MP4
109. Griffin, A., Nisi, K., Pepper, J., Harvey, A., Szydłowska, B.M., Coleman, J.N., Backes, C.: Effect of surfactant choice and concentration on the dimensions and yield of liquid-phase-exfoliated nanoparticles. Chem. Mater. **32**, 2852–2862 (2020). https://doi.org/10.1021/ACS.CHEMMATER.9B04684/ASSET/IMAGES/LARGE/CM9B04684_0006.JPEG
110. Saleh, H.M., Hassan, A.I.: Synthesis and characterization of nanomaterials for application in cost-effective electrochemical devices. Sustainability. **15**, 10891 (2023). https://doi.org/10.3390/SU151410891
111. Hu, R., Huang, Z., Wang, B., Qiao, H., Qi, X.: Electrochemical exfoliation of molybdenum disulfide nanoparticles for high-performance supercapacitors. J. Mater. Sci. Mater. Electron. **32**, 7237–7248 (2021). https://doi.org/10.1007/S10854-021-05432-5/METRICS
112. Wells, R.A., Zhang, M., Chen, T.H., Boureau, V., Caretti, M., Liu, Y., Yum, J.H., Johnson, H., Kinge, S., Radenovic, A., Sivula, K.: High performance semiconducting nanoparticles via a scalable powder-based electrochemical exfoliation technique. ACS Nano. **16**, 5719–5730 (2022). https://doi.org/10.1021/ACSNANO.1C10739/ASSET/IMAGES/MEDIUM/NN1C10739_M003.GIF
113. Yu, X., Peng, Z., Xu, L., Shi, W., Li, Z., Meng, X., He, X., Wang, Z., Duan, S., Tong, L., Huang, X., Miao, X., Hu, W., Ye, L.: Manipulating 2D materials through strain engineering. Small. **20**, 2402561 (2024). https://doi.org/10.1002/SMLL.202402561;CTYPE:STRING:JOURNAL
114. Fox, C., Mao, Y., Zhang, X., Wang, Y., Xiao, J.: Stacking order engineering of two-dimensional materials and device applications. Chem. Rev. **124**, 1862–1898 (2024). https://doi.org/10.1021/ACS.CHEMREV.3C00618/ASSET/IMAGES/LARGE/CR3C00618_0020.JPEG
115. García-Dalí, S., Paredes, J.I., Munuera, J.M., Villar-Rodil, S., Adawy, A., Martínez-Alonso, A., Tascón, J.M.D.: Aqueous cathodic exfoliation strategy toward solution-processable and phase-preserved MoS2 nanoparticles for energy storage and catalytic applications. ACS Appl. Mater. Interfaces. **11**, 36991–37003 (2019). https://doi.org/10.1021/ACSAMI.9B13484/ASSET/IMAGES/LARGE/AM9B13484_0008.JPEG

116. Martínez-Jódar, A., Villar-Rodil, S., Munuera, J.M., Castro-Muñiz, A., Coleman, J.N., Raymundo-Piñero, E., Paredes, J.I.: Two-dimensional MoS2 nanoparticles derived from cathodic exfoliation for lithium storage applications. Nanomaterials. **14**, 932 (2024). https://doi.org/10.3390/NANO14110932/S1
117. Liu, N., Kim, P., Kim, J.H., Ye, J.H., Kim, S., Lee, C.J.: Large-area atomically thin MoS2 nanoparticles prepared using electrochemical exfoliation. ACS Nano. **8**, 6902–6910 (2014). https://doi.org/10.1021/NN5016242/SUPPL_FILE/NN5016242_SI_001.PDF
118. Kim, S., Choi, M.S., Qu, D., Ra, C.H., Liu, X., Kim, M., Song, Y.J., Yoo, W.J.: Effects of plasma treatment on surface properties of ultrathin layered MoS2. 2D Mater. **3**, 035002 (2016). https://doi.org/10.1088/2053-1583/3/3/035002
119. Xiao, S., Xiao, P., Zhang, X., Yan, D., Gu, X., Qin, F., Ni, Z., Han, Z.J., Ostrikov, K.K.: Atomic-layer soft plasma etching of MoS2. Sci. Rep. **6**, 1–8 (2016). https://doi.org/10.1038/SREP19945; TECHMETA=128,133,140;SUBJMETA=1018,301,357,551,639,925;KWRD=SYNTHESIS+AND+PROCESSING,TWO-DIMENSIONAL+MATERIALS
120. Stark, M.S., Kuntz, K.L., Martens, S.J., Warren, S.C.: Intercalation of layered materials from bulk to 2D. Adv. Mater. **31**, 1808213 (2019). https://doi.org/10.1002/ADMA.201808213
121. Su, R., Zhu, X., Li, L., Dong, Z.: Research progress of MoS2-based electrode materials in electrocatalytic hydrogen evolution: from synthesis, characterization to electrocatalytic performance optimization strategy. Int. J. Hydrog. Energy. **137**, 26–72 (2025). https://doi.org/10.1016/J.IJHYDENE.2025.04.390
122. Shembade, U.V., Moholkar, R.A., Khot, R.S., Bhosale, T.T., Chougale, N.B., Magadum, M.G., Wategaonkar, S.B., Moholkar, A.V.: Future perspective of sustainable supercapacitors. In: Sustainable Supercapacitors: Next-Generation of Green Energy Storage Devices, pp. 323–352. Wiley, New York (2024). https://doi.org/10.1002/9781394237920.CH11
123. Vasi, S., Giofrè, S.V., Perathoner, S., Mallamace, D., Abate, S., Wanderlingh, U.: X-ray characterizations of exfoliated MoS2 produced by microwave-assisted liquid-phase exfoliation. Materials. **17**, 3887 (2024). https://doi.org/10.3390/MA17163887/S1
124. Hai, X., Chang, K., Pang, H., Li, M., Li, P., Liu, H., Shi, L., Ye, J.: Engineering the edges of MoS2 (WS2) crystals for direct exfoliation into monolayers in polar micromolecular solvents. J. Am. Chem. Soc. **138**, 14962–14969 (2016). https://doi.org/10.1021/JACS.6B08096/ASSET/IMAGES/JA-2016-080962_M001.GIF
125. Jawaid, A., Nepal, D., Park, K., Jespersen, M., Qualley, A., Mirau, P., Drummy, L.F., Vaia, R.A.: Mechanism for liquid phase exfoliation of MoS2. Chem. Mater. **28**, 337–348 (2016). https://doi.org/10.1021/ACS.CHEMMATER.5B04224/ASSET/IMAGES/LARGE/CM-2015-04224D_0007.JPEG
126. Jiang, F., Xiong, J., Zhou, W., Liu, C., Wang, L., Zhao, F., Liu, H., Xu, J.: Use of organic solvent-assisted exfoliated MoS2 for optimizing the thermoelectric performance of flexible PEDOT:PSS thin films. J. Mater. Chem. A Mater. **4**, 5265–5273 (2016). https://doi.org/10.1039/C6TA00305B
127. Kumar, S., Mishra, T.: Shock wave induced exfoliation of molybdenum disulfide (MoS2) in various solvents: all-atom molecular dynamics simulation. J. Mol. Liq. **314**, 113671 (2020). https://doi.org/10.1016/J.MOLLIQ.2020.113671
128. Zhang, H., Chen, X., Wang, J., Zhang, Y., Wang, G., Mao, Y., Wang, Z., Zhang, Y., Gu, R., Zhao, M.: Solvent-assisted exfoliation for high-quality molybdenum disulfide nanoflakes and relevant field-effect transistors. J. Mater. Sci. **57**, 11215–11225 (2022). https://doi.org/10.1007/S10853-022-07374-7/TABLES/2
129. Saima, N.A., Hossain, K.S., Hussain, S.: Preparation of MoS2 nanoflakes in a mixed solvent by liquid phase exfoliation (LPE) technique. AIP Adv. **14**, 125017 (2024). https://doi.org/10.1063/5.0235505/3325607
130. Wu, S., Lee, W.P.C., Thenuwara, H.N., Wu, P.: Quantitative criteria for solvent selection in liquid-phase exfoliation: balancing exfoliation and stabilization efficiency. Nanomaterials. **15**, 370 (2025). https://doi.org/10.3390/NANO15050370

131. Presolski, S., Pumera, M.: Covalent functionalization of MoS2. Mater. Today. **19**, 140–145 (2016). https://doi.org/10.1016/J.MATTOD.2015.08.019
132. Pourmadadi, M., Tajiki, A., Hosseini, S.M., Samadi, A., Abdouss, M., Daneshnia, S., Yazdian, F.: A comprehensive review of synthesis, structure, properties, and functionalization of MoS2; emphasis on drug delivery, photothermal therapy, and tissue engineering applications. J. Drug Deliv. Sci. Technol. **76**, 103767 (2022). https://doi.org/10.1016/J.JDDST.2022.103767
133. Mphuthi, N., Sikhwivhilu, L., Ray, S.S.: Functionalization of 2D MoS2 nanoparticles with various metal and metal oxide nanostructures: their properties and application in electrochemical sensors. Biosensors. **12**, 386 (2022). https://doi.org/10.3390/BIOS12060386
134. Zhang, J., Zhu, Y., Tebyetekerwa, M., Li, D., Liu, D., Lei, W., Wang, L., Zhang, Y., Lu, Y.: Vanadium-doped monolayer MoS2 with tunable optical properties for field-effect transistors. ACS Appl. Nano Mater. **4**, 769–777 (2021). https://doi.org/10.1021/ACSANM.0C03083/ASSET/IMAGES/LARGE/AN0C03083_0007.JPEG
135. Li, M., Cai, B., Tian, R., Yu, X., Breese, M.B.H., Chu, X., Han, Z., Li, S., Joshi, R., Vinu, A., Wan, T., Ao, Z., Yi, J., Chu, D.: Vanadium doped 1T MoS2 nanoparticles for highly efficient electrocatalytic hydrogen evolution in both acidic and alkaline solutions. Chem. Eng. J. **409**, 128158 (2021). https://doi.org/10.1016/J.CEJ.2020.128158
136. Shi, W., Wang, Z.: Titanium-doped MoS2 monolayer as highly efficient catalyst for hydrogen evolution reaction. IOP Conf. Ser. Earth Environ. Sci. **558**, 032048 (2020). https://doi.org/10.1088/1755-1315/558/3/032048
137. Tang, K., Wang, L., Geng, H., Qiu, J., Cao, H., Liu, X.: Molybdenum disulfide (MoS2) nanoparticles vertically coated on titanium for disinfection in the dark. Arab. J. Chem. **13**, 1612–1623 (2020). https://doi.org/10.1016/J.ARABJC.2017.12.013
138. Khan, M.I., Hasan, M.S., Bhatti, K.A., Rizvi, H., Wahab, A., Rehman, S.U., Afzal, M.J., Nazneen, A., Fiaz Khan, M., Nazir, A., Iqbal, M.: Effect of Ni doping on the structural, optical and photocatalytic activity of MoS2, prepared by hydrothermal method. Mater. Res. Express. **7**, 015061 (2020). https://doi.org/10.1088/2053-1591/AB66F7
139. Romero Garcia, S., Faiyad, A., Martini, A.: Modeling sputtering deposition of MoS2: effect of Ni doping on nanostructure and tribological properties. Comput. Mater. Sci. **244**, 113229 (2024). https://doi.org/10.1016/J.COMMATSCI.2024.113229
140. Alharthi, F.A., Hasan, I.: Visible light-driven efficient photocatalytic hydrogen production using nickel-doped molybdenum disulfide (Ni@MoS2) nanoflowers. J. Mater. Sci. **59**, 3394–3405 (2024). https://doi.org/10.1007/S10853-024-09435-5/TABLES/1
141. Sharma, C., Srivastava, A.K., Gupta, M.K.: High performance two dimensional few layered copper doped MoS2 nanoparticles based flexible piezoelectric nanogenerator. Indian J. Pure Appl. Phys. (IJPAP). **62**, 102–108 (2024). https://doi.org/10.56042/IJPAP.V62I2.7678
142. Hussain, A., Tayyab, M.: Effect of Cu concentration and dopant site on the band gap of MoS2: a DFT study. Comput. Condens. Matter. **24**, e00494 (2020). https://doi.org/10.1016/J.COCOM.2020.E00494
143. Luo, Z., Ouyang, Y., Zhang, H., Xiao, M., Ge, J., Jiang, Z., Wang, J., Tang, D., Cao, X., Liu, C., Xing, W.: Chemically activating MoS2 via spontaneous atomic palladium interfacial doping towards efficient hydrogen evolution. Nat. Commun. **9**, 1–8 (2018). https://doi.org/10.1038/s41467-018-04501-4
144. Hu, A., Shu, C., Xu, C., Li, J., Liang, R., Zheng, R., Li, M., Ran, Z., Long, J.: Interface-engineered metallic 1T-MoS2 nanosheet array induced via palladium doping enabling catalysis enhancement for lithium–oxygen battery. Chem. Eng. J. **382**, 122854 (2020). https://doi.org/10.1016/J.CEJ.2019.122854
145. Li, Y., Gu, Q., Johannessen, B., Zheng, Z., Li, C., Luo, Y., Zhang, Z., Zhang, Q., Fan, H., Luo, W., Liu, B., Dou, S., Liu, H.: Synergistic Pt doping and phase conversion engineering in two-dimensional MoS2 for efficient hydrogen evolution. Nano Energy. **84**, 105898 (2021). https://doi.org/10.1016/J.NANOEN.2021.105898
146. Li, H., Wang, S., Sawada, H., Han, G.G.D., Samuels, T., Allen, C.S., Kirkland, A.I., Grossman, J.C., Warner, J.H.: Atomic structure and dynamics of single platinum atom inter-

actions with monolayer MoS2. ACS Nano. **11**, 3392–3403 (2017). https://doi.org/10.1021/ACSNANO.7B00796/ASSET/IMAGES/NN-2017-00796A_M001.GIF
147. Akhtar, P., Khan, M.J.I., Ramay, S.M., Mahmood, A., Saleem, M., Kanwal, Z.: Ab-initio and experimental investigations on Pt:MoS2 for electronic and optical applications. Chem. Phys. Lett. **780**, 138938 (2021). https://doi.org/10.1016/J.CPLETT.2021.138938
148. Guan, T., Zhang, N.: Recent advances in electrodeposition of nickel-based nanocomposites enhanced with lubricating nanoparticles. Nanomanuf. Metrol. **7**, 1–38 (2024). https://doi.org/10.1007/S41871-024-00245-6
149. George, C., Morris, A.J., Modarres, M.H., De Volder, M.: Structural evolution of electrochemically lithiated MoS2 nanoparticles and the role of carbon additive in Li-ion batteries. Chem. Mater. **28**, 7304–7310 (2016). https://doi.org/10.1021/ACS.CHEMMATER.6B02607/ASSET/IMAGES/LARGE/CM-2016-02607Y_0003.JPEG
150. Liu, Q., Huo, J., Ma, Z., Wu, Z., Wang, S.: In-situ formation of Ni3S2 interlayer between MoS2 and Ni foam for high-rate and highly-durable lithium ion batteries. Electrochim. Acta. **206**, 52–60 (2016). https://doi.org/10.1016/J.ELECTACTA.2016.04.121
151. Han, J., Shao, Z., Jia, Y., Liu, P., Han, Z., Zhang, X., Wu, L., Ma, X.: Hierarchical NiS2/1T-MoS2 nanocomposite for stable lithium sulfur batteries. ACS Appl. Nano Mater. **7**(22), 25637–25644 (2024). https://doi.org/10.1021/ACSANM.4C04738/SUPPL_FILE/AN4C04738_SI_001.PDF
152. Shokhen, V., Zitoun, D.: Platinum-group metal grown on vertically aligned MoS2 as electrocatalysts for hydrogen evolution reaction. Electrochim. Acta. **257**, 49–55 (2017). https://doi.org/10.1016/J.ELECTACTA.2017.10.014
153. Joseph, K.S., Dabhi, S.: Transition metal embedded MoS2 as efficient trifunctional electrocatalyst: a computational study. Electrochim. Acta. **505**, 144968 (2024). https://doi.org/10.1016/J.ELECTACTA.2024.144968
154. Kumari, S., Raturi, S., Kulshrestha, S., Chauhan, K., Dhingra, S., András, K., Thu, K., Khargotra, R., Singh, T.: A comprehensive review on various techniques used for synthesizing nanoparticles. J. Mater. Res. Technol. **27**, 1739–1763 (2023). https://doi.org/10.1016/J.JMRT.2023.09.291
155. Saraloğlu Güler, E., Konca, E., Karakaya, I.: Effect of electrodeposition parameters on the current density of hydrogen evolution reaction in Ni and Ni-MoS2 composite coatings. Int. J. Electrochem. Sci. **8**, 5496–5505 (2013). https://doi.org/10.1016/S1452-3981(23)14699-2
156. Giang, H., Adil, O., Suni, I.I.: Electrodeposition of Ni-doped MoS2 thin films. J. Electrochem. Soc. **167**, 082512 (2020). https://doi.org/10.1149/1945-7111/AB8CE0
157. Hostert, L., Dias, M.S., de Aquino, C.B., dos Santos, F.C., Marangoni, V.S., de Carvalho, C., Silva, C., Seixas, L., Maroneze, C.M.: Covalent surface functionalization of exfoliated MoS2 nanoparticles for improved electrocatalysis. J. Phys. Chem. C. **128**, 20856–20865 (2024). https://doi.org/10.1021/ACS.JPCC.4C05965/ASSET/IMAGES/LARGE/JP4C05965_0005.JPEG
158. Zhang, J., Zhao, X., Yang, Y., Cui, J.: Monolayer MoS2 with S vacancy defects doped with group V non-metallic elements (N, P, As): a first-principles study. J. Mol. Model. **31**, 64 (2025). https://doi.org/10.1007/S00894-025-06290-2/FIGURES/7
159. Peng, C., Song, L., Wang, L., Yang, F., Ding, J., Huang, F., Wang, Y.: Effect of surface charge distribution of phosphorus-doped MoS2 on hydrogen evolution reaction. ACS Appl. Energy Mater. **4**, 4887–4896 (2021). https://doi.org/10.1021/ACSAEM.1C00485/SUPPL_FILE/AE1C00485_SI_001.PDF
160. Hirsch, A., Hauke, F.: Post-graphene 2D chemistry: the emerging field of molybdenum disulfide and black phosphorus functionalization. Angew. Chem. Int. Ed. **57**, 4338–4354 (2018). https://doi.org/10.1002/ANIE.201708211;WGROUP:STRING:PUBLICATION
161. Jeon, S., Kim, M., Jia, J., Park, J.H., Lee, S., Song, Y.J.: Controlled p-doping of black phosphorus by integration of MoS2 nanoparticles. Appl. Surf. Sci. **440**, 282–287 (2018). https://doi.org/10.1016/J.APSUSC.2018.01.141

162. Sattari, S., Beyranvand, S., Soleimani, K., Rossoli, K., Salahi, P., Donskyi, I.S., Donskyi, I.S., Shams, A., Unger, W.E.S., Yari, A., Farjanikish, G., Nayebzadeh, H., Adeli, M.: Boronic acid-functionalized two-dimensional MoS2 at biointerfaces. Langmuir. **36**, 6706–6715 (2020). https://doi.org/10.1021/ACS.LANGMUIR.0C00776/SUPPL_FILE/LA0C00776_SI_001.PDF
163. Routh, P.: Synthesis of oxygen-rich edge functionalized MoS2 and WS2 quantum dots for cell imaging. ChemistrySelect. **10**, e202405165 (2025). https://doi.org/10.1002/SLCT.202405165;WGROUP:STRING:PUBLICATION
164. Tang, J., Wei, Z., Wang, Q., Wang, Y., Han, B., Li, X., Huang, B., Liao, M., Liu, J., Li, N., Zhao, Y., Shen, C., Guo, Y., Bai, X., Gao, P., Yang, W., Chen, L., Wu, K., Yang, R., Shi, D., Zhang, G.: In situ oxygen doping of monolayer MoS2 for novel electronics. Small. **16**, 2004276 (2020). https://doi.org/10.1002/SMLL.202004276;WGROUP:STRING:PUBLICATION
165. Im, H., Bala, A., So, B., Kim, Y.J., Kim, S.: Customization of MoS2 phototransistors via thiol-based functionalization. Adv. Electron. Mater. **7**, 2100644 (2021). https://doi.org/10.1002/AELM.202100644
166. Jia, Y., Zhang, Y., Xu, H., Li, J., Gao, M., Yang, X.: Recent advances in doping strategies to improve electrocatalytic hydrogen evolution performance of molybdenum disulfide. ACS Catal. **14**, 4601–4637 (2024). https://doi.org/10.1021/ACSCATAL.3C05053
167. Kang, M., Hong, W., Lee, I., Park, S., Park, C., Bae, S., Lim, H., Choi, S.Y.: Tunable doping strategy for few-layer MoS2 field-effect transistors via NH3 plasma treatment. ACS Appl. Mater. Interfaces. **16**, 43849–43859 (2024). https://doi.org/10.1021/ACSAMI.4C08549
168. Le, K., Zhang, X., Zhao, Q., Liu, Y., Yi, P., Xu, S., Liu, W.: Controllably doping nitrogen into 1T/2H MoS2 heterostructure nanoparticles for enhanced supercapacitive and electrocatalytic performance by low-power N2 plasma. ACS Appl. Mater. Interfaces. **13**, 44427–44439 (2021). https://doi.org/10.1021/ACSAMI.1C12973
169. Yang, Q., Wang, Z., Dong, L., Zhao, W., Jin, Y., Fang, L., Hu, B., Dong, M.: Activating MoS2 with super-high nitrogen-doping concentration as efficient catalyst for hydrogen evolution reaction. J. Phys. Chem. C. **123**, 10917–10925 (2019). https://doi.org/10.1021/ACS.JPCC.9B00059/SUPPL_FILE/JP9B00059_SI_001.PDF
170. Saha, D., Barakat, S., Van Bramer, S.E., Nelson, K.A., Hensley, D.K., Chen, J.: Noncompetitive and competitive adsorption of heavy metals in sulfur-functionalized ordered mesoporous carbon. ACS Appl. Mater. Interfaces. **8**, 34132–34142 (2016). https://doi.org/10.1021/ACSAMI.6B12190/ASSET/IMAGES/LARGE/AM-2016-12190D_0011.JPEG
171. Mondal, A., De, M.: Amino acid-functionalized MoS2 quantum dots for selective antibacterial activity. ACS Appl. Nano Mater. **4**, 13947–13954 (2021). https://doi.org/10.1021/ACSANM.1C03243/SUPPL_FILE/AN1C03243_SI_001.PDF
172. Chiu, N.F., Lin, T.L.: Affinity capture surface carboxyl-functionalized MoS2 sheets to enhance the sensitivity of surface plasmon resonance immunosensors. Talanta. **185**, 174–181 (2018). https://doi.org/10.1016/J.TALANTA.2018.03.073
173. Wang, L., Wu, J., Wang, X., Fu, S.: Defect engineering of MoS2-based materials as supercapacitors electrode: a mini review. J. Alloys Compd. **959**, 170548 (2023). https://doi.org/10.1016/J.JALLCOM.2023.170548
174. Cho, K., Min, M., Kim, T.Y., Jeong, H., Pak, J., Kim, J.K., Jang, J., Yun, S.J., Lee, Y.H., Hong, W.K., Lee, T.: Electrical and optical characterization of MoS2 with sulfur vacancy passivation by treatment with alkanethiol molecules. ACS Nano. **9**, 8044–8053 (2015). https://doi.org/10.1021/ACSNANO.5B04400/ASSET/IMAGES/LARGE/NN-2015-04400S_0009.JPEG
175. Tsai, C., Li, H., Park, S., Park, J., Han, H.S., Nørskov, J.K., Zheng, X., Abild-Pedersen, F.: Electrochemical generation of sulfur vacancies in the basal plane of MoS2 for hydrogen evolution. Nat. Commun. **8**, 1–8 (2017). https://doi.org/10.1038/ncomms15113
176. Hu, J., Yu, L., Deng, J., Wang, Y., Cheng, K., Ma, C., Zhang, Q., Wen, W., Yu, S., Pan, Y., Yang, J., Ma, H., Qi, F., Wang, Y., Zheng, Y., Chen, M., Huang, R., Zhang, S., Zhao, Z., Mao,

J., Meng, X., Ji, Q., Hou, G., Han, X., Bao, X., Wang, Y., Deng, D.: Sulfur vacancy-rich MoS2 as a catalyst for the hydrogenation of CO2 to methanol. Nat. Catal. **4**, 242–250 (2021). https://doi.org/10.1038/s41929-021-00584-3
177. Mondal, S., Dilly Rajan, K., Patra, L., Rathinam, M., Ganesh, V.: Sulfur vacancy-induced enhancement of piezocatalytic H2 production in MoS2. Small. **21**, 2411828 (2025). https://doi.org/10.1002/SMLL.202411828; PAGE:STRING:ARTICLE/CHAPTER
178. Lu, W., Birmingham, B., Zhang, Z.: Defect engineering on MoS2 surface with argon ion bombardments and thermal annealing. Appl. Surf. Sci. **532**, 147461 (2020). https://doi.org/10.1016/J.APSUSC.2020.147461
179. Donarelli, M., Bisti, F., Perrozzi, F., Ottaviano, L.: Tunable sulfur desorption in exfoliated MoS2 by means of thermal annealing in ultra-high vacuum. Chem. Phys. Lett. **588**, 198–202 (2013). https://doi.org/10.1016/J.CPLETT.2013.10.034
180. Luo, C.W., Cai, L., Xie, C., Li, G., Jiang, T.J.: Sulfur vacancies on MoS2 enhanced the activation of peroxymonosulfate through the co-existence of radical and non-radical pathways to degrade organic pollutants in wastewater. RSC Adv. **12**, 25364–25376 (2022). https://doi.org/10.1039/D2RA02448A
181. Yin, Y., Han, J., Zhang, Y., Zhang, X., Xu, P., Yuan, Q., Samad, L., Wang, X., Wang, Y., Zhang, Z., Zhang, P., Cao, X., Song, B., Jin, S.: Contributions of phase, sulfur vacancies, and edges to the hydrogen evolution reaction catalytic activity of porous molybdenum disulfide nanoparticles. J. Am. Chem. Soc. **138**, 7965–7972 (2016). https://doi.org/10.1021/JACS.6B03714/ASSET/IMAGES/LARGE/JA-2016-03714Y_0006.JPEG
182. Jeong, M., Kim, S., Ju, S.Y.: Preparation and characterization of a covalent edge-functionalized lipoic acid–MoS2 conjugate. RSC Adv. **6**, 36248–36255 (2016). https://doi.org/10.1039/C6RA00901H
183. Wypych, F., Weber, T., Prins, R.: Scanning tunneling microscopic investigation of 1T-MoS2. Chem. Mater. **10**, 723–727 (1998). https://doi.org/10.1021/CM970402E/ASSET/IMAGES/LARGE/CM970402EF00006.JPEG
184. Acerce, M., Voiry, D., Chhowalla, M.: Metallic 1T phase MoS2 nanoparticles as supercapacitor electrode materials. Nat. Nanotechnol. **4**(10), 313–318 (2015). https://doi.org/10.1038/nnano.2015.40
185. Py, M.A., Haering, R.R.: Structural destabilization induced by lithium intercalation in MoS2 and related compounds. Can. J. Phys. **61**, 76–84 (2011). https://doi.org/10.1139/P83-013
186. Awate, S.S., Xu, K., Liang, J., Katz, B., Muzzio, R., Crespi, V.H., Katoch, J., Fullerton-Shirey, S.K.: Strain-induced 2H to 1T′ phase transition in suspended MoTe2 using electric double layer gating. ACS Nano. **17**, 22388–22398 (2023). https://doi.org/10.1021/ACSNANO.3C04701/ASSET/IMAGES/LARGE/NN3C04701_0006.JPEG
187. Taft, C.A., de Lazaro, S.R. (eds.): Progress in Hydrogen Energy Fuel Cells, Nano-Biotechnology and Advanced, Bioactive Compounds. Springer Nature, Cham (2024). https://doi.org/10.1007/978-3-031-75984

Overview: Multidisciplinary Applications of Quantum Dots in Energy, Biomedicine and Other Advanced Technologies

Carlton A. Taft and Jose Gabriel Solano Canchaya

Abstract We overview quantum dots (QDs), (2023 Nobel Prize in Chemistry), underlining the many areas of important ***MULTIDISCIPLINARY APPLICATIONS*** in progress, with emphasis on energy, biomedicine, quantum computing and other selected advanced technologies. QDs, nickname artificial atoms, are cutting-edge technology, zero dimensional (0D), semiconductor nanocrystals (2–10 nm diameter), behaving like single atoms, smaller than *Bohr exciton radius*, consequently with properties different from bulk (governed by quantum mechanics), i.e. quantum confinement (energy states discretization) allowing extraordinary size dependent properties such as tunable fluorescence wavelengths and band gaps. Numerous QDs properties can be further modified by external manipulation, offering a wide range of important technological/industrial applications and innovative approaches to visualization, diagnosis, treatment and disease prevention. QDs comprise a wide range of structures/materials/properties and can be classified based on the positioning in the periodic table, typically including group II–VI, III–V, IV–VI, IV, 2D QDs, carbon/graphene QDs and perovskite QDs. Applications (many in our daily lives) include (medical diagnostics, monitoring, treatment, tracking, bioimaging, biosensing, gene therapy/delivery, bioanalytical detection, nanomedicine, intracellular/traceable drug delivery for (cancer therapy (skin, head, neck, tongue, pancreatic, breast, hepatocellular carcinoma, ...), brain disease, gastric ulcer, liver fibrosis, ulcerative colitis, diabetes, cardiovascular disease, amylin-related diseases, leukemia, dental applications, angiogenesis, antimicrobial activity, multidrug-resistant bacteria, wound healing, Alzheimer's disease, Parkinson's disease, immunological diagnosis); DNA-functionalized probes, pH sensing, pharmaceutical formulations, biomedical/tissue engineering, targeted surgery, labelling, new generation implants, early-stage diag-

C. A. Taft (✉)
Brazilian Center for Physics Research, Rio de Janeiro, Brazil

J. G. S. Canchaya
Graduate School, Universidad Continental, Lima, Peru

Group of Computational Methods Applied to Nanomaterials, Nacional University of San Marcos, Lima, Peru

C. A. Taft, S. R. d. Lazaro (eds.), *Innovative Materials for Energy and Sustainability*, Engineering Materials,
https://doi.org/10.1007/978-3-032-09149-9_2

nosis, photodynamic therapy, radiotherapy); fuel cells, solar cells, energy storage devices, solar energy harvesting, catalysts, waste water treatment, photocatalytic hydrogen synthesis, microplastic degradation, environmental remediation, pollutant detection and removal, agriculture, seed germination, organic dye degradation, nitric oxide oxidation; batteries (lithium-sulfur, zinc-air, primary, rechargeable), supercapacitors, sensors, light emitting diodes, lasers, photonics, quantum computing/information/sensing, AI material optimization, QD-miniled/beyond OLED, QDEL displays, wearable electronics, cameras, machine vision, bioactive glasses, photoconductors, photodetectors, phototransistors, screens, antennas; consumer goods including (scratch-resistant surface coatings, cosmetics, inkjet printing, anti-fake detection, perfumes, microwaves, tools, paints, flame retardancy, forensic techniques, fire hazards, thermal stability, sports goods, fingerprints detection).

1 Introduction

Quantum Dots (QDs) are zero-dimensional nanomaterial semiconductors, diameters 2–10 nm, (smaller 10,000,000 times than a tennis ball, smaller than Bohr electron wavelength), single point, confined length/breadth/heights, consisting typically of elements from groups II–VI, III–V, IV–VI, IV as well as carbon, graphite, perovskite…, using core, core-shell and alloyed structures (allowing participation of wide range of atoms), which can be further modified/controlled by external manipulation (magnetic fields, electric fields, acoustic waves, temperature, doping, and surface modification), with unique quantum confinement (variable number of electrons in well-defined/restricted location), indicating size/shape/structure-dependent/optical tunable properties, first described in the 1980s, are denoted as quantum dots (QDs), indicating remarkable combination of adjustable *quantum behavior/optoelectronic properties*, attracting thereof considerable research interest (rocketing publication increase in the last few decades) due to their roles such as light emission, conversion, detection and consequent wide range applicability including biomedical, technological areas and consumer product versatility [1–40].

The quantum confinement of QDs can in principle be rationalized by quantum mechanics foundational models, i.e., 'particle in a box' whereas the nanocrystal's energy levels (responsible for optical properties) increases with reduction in box size and the surface defines the box's boundary [39]. In nanoforms, quantum confinement induces discrete energy levels which contribute to unique QDs optical characteristics. When light is absorbed in bulk materials electrons and holes can move independently (exhibiting a spectrum of "continuous" energy levels) whereas for QDs the small size affords 'discrete energy levels', energy confinement (electron-hole (e-h) pairs compression), within nanocrystals grain boundaries. This confinement yields unique visual characteristics/photo-physical features which encompass, among others narrow tunable emission, photochemical durability, broad absorption spectra, intermittency, whereas fluorescence can be controlled over 400–4000 nm wavelength range enabling near-infrared, visible, ultraviolet spectra measurements as well as Stokes shifts (good separation between longer-wavelength emission spectra and shorter-wavelength excitation spectra) with consequent improvement of reduction in optical overlap and detection sensitivity) [1–39].

The 3D quantum confinement (physically confined electron cloud produced by the minute size of the QDs) governs dimensions, discrete energy levels, size-dependent properties which enables a wide range of functionalities and tunability. Most QDs can emit specific light wavelengths whereas electrons become excited to higher energy states with release of photons upon return to ground state. Since the wavelength of emissions are size controllable the electronic characteristics as well as color of emitted photon is also size dependent yielding for example at shorter wavelengths (blue, green) for relatively shorter radius range size and longer wavelengths closer to (red, orange) for higher radius range size. The light color can be controlled by controlling QD size. Optical performance is more size-dependent than material-dependent allowing electromagnetic radiation manipulation [9–60].

The QDs unique blinking features, unlike conventional probes, makes easier single events detections offering capability of for example individual subcellular components visualization such as proteins. The emission wavelength is reportedly proportional to particle diameter whereas particle size/emission spectra can be manipulated from core composition. They also indicate other interesting features such as high quantum yield, brightness, photostability, water solubility, high tissue accumulation, ease of fabrication and capability of conjugation to a wide variety of drugs [9–70].

QDs address a wide range of structures, properties, materials including (1) group II–VI (CdSe, CdS, ZnS, ZnO, ZnTe, ZnSe, HgTe), group III–V (InP, InN, InSb, InAs, GaP, GaN, GaSb, GaAs, AlP, AlN, AlSb), group IV–VI (PbSe, PbS, PTe, PbS), group IV(Ge,Si); (2) 2D QDs (transition metal dichalcogenide (TMD) QDs, graphene QDs (GQDs), graphitic carbon nitride (g-C_3N_4) or QDs (g-CNQDs), hexagonal boron nitride or (H-BN) QDs, black phosphorus QDs (BP QDs), silicene, phosphorene, MXenes); (3) carbon QDs or (CQDs); (4) perovskite QDs. Quantum dots types include core–type QDs (QDs), (single-component materials); core–shell QDs (CSQDs), i.e., small regions of one material embedded with another bandgap material; alloyed QDs (QDs formed by two nanocrystal semiconductors). This wide variety of QDs yields important properties with great/wide applicability and yet much more future potential for numerous applications in multidisciplinary fields and industries [9–80].

QDs preparation methods are typically divided into top-down and bottom-up methods. A bulk semiconductor is transformed into QDs in the top-down methods with a typical diameter of 30 nm. This model is very effective for investigating novel materials, their characteristics and structures whereas methodologies such as electron beam lithography, dry focus beam method and wet chemical etching are mostly used. The bottom-up methodology (vapor-phase, wet-chemical) starts with basic materials and produces high-quality nanomaterials with precise size, superior homogeneity, enhanced reproducibility and fewer defects [13–50].

QDs offer broad optical range of fluorescence with narrow emission spectra and large Stokes shifts which have greatly expanded the potential of fluorescence in biomedicine. For example one could inject various sizes QDs into patients which when excited with single wavelengths would result in a narrow emission (minimum cross-talk between channels, spectral overlaps) as well as single wavelengths which would yield improved sensitivity, enhanced detection efficiency, minimum cross-talk between channels enabling multi-modal functionality and multi-color detection. Notable it is challenging to modulate/control spectral wavelengths of traditional organic dyes (cannot be easily modified to emit different colors) imposing limitations on their application in multiplex imaging applications [1–30].

Another area of interest includes the exceptional stability and strength of GQDs which makes them important in oral environment where subjected to constant mechanical stress and whereas their dental materials can withstand abrasive actions, biting forces indicating as well the required biocompatibility and antimicrobial properties [9–30].

Quantum dots in quantum computing is of interest. Other areas of applicability are addressed in this overview. Actually, there is an active forefront/state of the art research using QDs.

Effectively, multidisciplinary wide range applications of QDs includes artificial intelligence (AI) material optimization, fuel cells, solar cells, zinc-air batteries, energy storage devices, catalysts, waste water treatment, photocatalytic hydrogen synthesis, microplastic degradation, environmental remediation, organic dye degradation, nitric oxide oxidation, batteries (lithium-sulfur, primary batteries, rechargeable), supercapacitors, sensors, light emitting diodes, lasers, infrared photodetector; biomedicine (medical diagnostics, monitoring, treatment, tracking, bioimaging, biosensing, gene therapy, bioanalytical detection, nanomedicine, cell tracking, intracellular/traceable drug delivery for including (cancer therapy (skin, head and neck, tongue, pancreatic, breast, hepatocellular carcinoma), brain disease, gastric ulcer, liver fibrosis, anxiety, blood glucose level, ulcerative colitis, diabetes, amylin-related diseases, leukemia, dental applications, antimicrobial activity, multidrug-resistant bacteria, toxicological, wound healing, Alzheimer's disease, Parkinson's disease, prion disease, anxiety disorder, immunological diagnosis), DNA-functionalized probes, pharmaceutical formulations, biomedical/tissue engineering, labelling, new generation implants, early-stage diagnosis/therapy, photodynamic therapy) [9–30]; consumer goods including (scratch-resistant surface coatings, cosmetics, inkjet printing, anti-fake detection, perfumes, microwaves, tools, paints, flame retardancy, forensic techniques, fire hazards, thermal stability, sports goods, fingerprints detection), photonics, quantum computing/information/sensing; displays, wearable electronics, cameras, photoconductors, photodetectors, phototransistors, screens, antennas [1–8, 31–126].

2 Biomedical Applications

In early biological work (gene therapy, photothermal treatment, sensing, imaging, drug travel) the nanomaterials were produced/applied in fields such as gold NPs for imaging, magnetic NPs for sensing/imaging and liposomes for drug delivery. In

1998 was first used quantum dots as biological labels, i.e. protein conjugation, biotin functionalization enabling intracellular imaging, cellular uptake, specific binding [9–30].

QDs can be conjugated to a number of biomolecules, including proteins, oligonucleotides, antibodies, and small molecule ligands, to direct their in vitro or in vivo pathway to a specific target or to impart therapeutic capability and today are important growing multidisciplinary area of Biomedicine [9–30] .

2.1 *Biomedical Carbon Quantum Dots (CQDs)*

Some semiconductor QDs although having excellent bioimaging and biosensing purposes, have inherent drawbacks including manufacturing heavy metal inexpensive toxicity. As a result, CQDs (discovered while isolating/purifying SWCNTs) were further developed as partial substitution for semiconductor quantum dots since they exhibit comparable fluorescence characteristics, are compatible with living organisms, chemically stable and less hazardous making them valuable for bioimaging and biosensing [9–30].

The CQDs catalytic/physicochemical features makes them biomedical application candidates with important features, including biocompatibility, tiny size allowing their usage as drug delivery vehicles as well as monitors in body due to PL qualities. Their chemical stability, hydrophilicity, low toxicity, wavelength-dependent PL emission, diverse sources of raw materials, simple surface functionalization, water solubility affords potential usage in the biomedical field.

2.2 *Bioimaging*

In research/clinical applications the biological reactions (targeted delivery, biological distribution, cellular uptake) can be analyzed in a precise, isolated form via the electromagnetic spectrum (multiple wavelengths). Researchers were thus prompted to develop sensitive, specific, low toxic graphene quantum dots GQD, stimulated by the fact that early diagnosis improved survival of patients. The excellent luminous character of GQD, for example, performing well in disease identification, imaging, makes them a good alternative to organic dyes whereas early tumor identification, disease recurrence, metastasis from sensitive imaging play important roles in cancer diagnostics [9–30].

In the biological field fluorescence imaging is essential to examine material distribution in tissues, organs, cells, using near-infrared spectra and visible light. Although dyes have been used as fluorescent probes, the water solubility, low toxicity of QDs have made them preferable [9–30].

2.3 In Vivo and Live Cell Imaging

The features of QDs have led to their usage as fluorescent probes for diversified biomedical imaging including visualization of the intracellular components whereas due to their fine particle size they can be taken up by cells and detected. Subsequently, they can be excited and the emission spectra detected by confocal laser scanning microscopy (CLSM) or fluorescence microscopes. Their blinking feature facilitates single event detection as well as individual subcellular component visualization. Their affinity to organs/tissues of interest can be increased by functionalization with selected ligands [9–30].

2.4 Magnetic Resonance Imaging

The spins of body protons are modified in MRI (magnetic resonance imaging) generating thus images for studying physiological processes. This noninvasive clinical imaging technique is able to penetrate deep tissues with high resolution whereas contrast agents can also be employed to enhance features. In some aggressive tumors (breast cancer, lung), silent micro metastasis yields fewer cancer cells circulating in the blood with difficult detection from regular methods. This problem can be addressed by using external magnetic fields, antibody-labelled QDs and detected by FM (Fluorescence Microscope) [9–30].

2.5 Cellular Imaging

When GQDs are absorbed by cells, cellular imaging can occur depending on functionalization and characteristics of the cellular compartments, whereas within the cells they can relocate to other cellular compartments. After decoration with specific targeting moieties, they can be used for intracellular visualization via fluorescent labels [9–30].

2.6 Fluorescence-Activated Cell Sorting (FACS)

FACS can be used for isolation of distinct cell populations, mapping of immune cells, evaluation of the cellular uptake of drug delivery systems, characterization of certain diseases models and detection of cellular markers [9–30].

QDs can be used in FACS since their narrow emission spectra reduces overlapping and multiple labels can be used for polychromatic cell sorting. The recruitment of single laser beams is facilitated by the broad excitation spectra leading to desired excitation of multiple QDs probes in FACS equipment.

Their brightness increases detection accuracy. Due to their price and high stability, they can replace antibodies that stain cell surface markers with objectives such as targeting ligands/recognizing target markers. The need for permeabilization buffers that increases difficulty of FACS experiments can be avoided.

2.7 Cancer Therapy

QDs surfaces can be modified for cancer imaging to examine tumor habitats, blood vessels, cells interconnected, encompassing the malignant growth whereas they can be used advantageously for glioblastoma, breast, prostate and other cancers. Tumors interact with environments where blood vessels supply the nutrients and can discharge into bloodstreams the cancer cells. QDs can integrate the medicine navigating blood arteries in tumors, targeting malignant cells and can be administered in different sizes and colors in order to investigate blood arteries diameters [9–30].

2.8 Cancer and Photodynamic Therapy (PDT)

PDT is used for treatment of cancers, including head, neck, tongue and skin cancers whereas a chemical 'photosensitizer' compound activated by light transfers energy to molecular oxygen (intracellular) resulting in reactive oxygen species (ROS) and apoptosis in tumor cells. QDs can be optimized to the NIR region to attain strong tissue penetration of deep-sited tumors [9–30].

2.9 Biosensors and Medical Applications

In response to biological processes, biosensors produce signals, which are measurable, depending on selectivity whereas bio responsive polymers, artificial receptors, antibodies, ligands or click chemistry are used. Efficiency, accuracy, detection sensitivity and selectivity can be incorporated into a single QD system and applied to diversified toxicological/diagnostic/medical applications [9–30].

2.9.1 Nitrogen Doped Graphene Quantum Dot Sensors

Nitrogen doped graphene quantum dots (N-GQDs) have attracted significant interest due to their important luminescence activity and incredible electrocatalytic stability in sensors with applications including electrochemiluminescence signal-on techniques [9–30].

N-GQDs have indicated great adhesion for loading to the electrode surface, i.e. hydrophilic properties, high water permeability, acted as activated catalyst, shown superior optical properties, increasing trends of fluorescence intensity, bright green luminescence, 1.5 times higher PL intensity than the bare GQDs, QY of 30.7%, bright blue and green luminescence, increasing fluorescence intensity with, high detection of dopamine sensitivity and enhancement in ammonia sensitivity.

2.10 Traceable Drug Delivery

The possibility of conjugation to a wide variety of drugs and special optical, physical-chemical properties makes QDs, after administration, traceable drug carriers whereas the nanocarriers ultra-small size is an important feature for penetrating the microenvironment of challenging cancer tumors. Notable, CQDs can also be used as biosensors for blood glucose level measurements [9–30].

Decoration with targeting ligands, physical chemical modifications, functional coatings can be used to modify pharmacokinetic properties and biodistribution where also retention time in the blood circulation is improved after intravenous administrations to achieve necessary tumor accumulation for cancer therapy. Functionalized QDs can be taken up by cells, whereas there are various pathways for the cells to react with ligand-modified QDs. Hybrid biosensor systems QDs, including graphene and multi-walled carbon nanotubes can be used to monitor processes involving/including therapeutic antibiotics, detection of lung cancer micro metastases in the blood and toxins in biological fluids.

2.11 Composite-Based CQDs for Sensing

The interaction of CQds with other nanomaterials such as Ag, Au can yield important new properties, including high surface area, microporosity. Good examples are properties obtained from interaction between organic frameworks (MOFs) and these QDs such as *sensing* as well as photocatalysis, storage, optoelectronics and electronic devices discussed in subsequent sections [9–30].

2.12 Optical Imaging

Nanoparticles can be used to identify molecular biomarkers and cancer cells as they are linked with antibodies, small molecules, peptides where QDs can be used for molecular diagnostics, bioimaging whereas single particle tracking (SPT) can be used for system statistical analysis via monitoring of individual molecule dynamics [9–30].

The optical characteristics of CQDs include high quantum yield, narrow emission spectra, wide absorption enabling their usage in bioimaging whereas quantum confinement yields size-based luminosity. Based on different sizes, CQDS with consistent chemical compositions can show varying wavelengths fluorescent emissions.

Excellent cellular and molecular mapping can be obtained from multiplexed conjugates of CQD antibodies serving as single-particle tracking at important/numerous levels such as brain slices, cell membrane and intracellular. It is sometimes necessary to apply additional coatings/shells in order to increase to 10–30 nm diameters of CQDs for better stabilization, biocompatibility, functionalization.

CQDs are excellent candidates for tissue/cell Imaging since they indicate water solubility, physicochemical stability, nontoxicity, biocompatibility, nonblinking features. They can be used to directly target malignant cells by navigating in tumors permeable blood arteries resulting in medications.

2.13 Antimicrobial Activity

CQDs have indicated to be efficient regarding interaction with some viruses and some infection inhibitions including efficacy against herpes simplex virus type I as well as inhibition of the entry of human coronavirus-229E. They have also been employed as agents against antibacterial strains such as Staphylococcus aureus, E. coli and pseudomonas aeruginosa by visualizing of these microorganisms. Effectively, the presence of boronic acid or amino groups on CQDs can affect entry of simplex virus type 1 herpes [9–30].

2.14 Core Shell Architecture for Biomedical Applications

Modification of CDs towards a core-shell structure using molecular imprinted polymers as shell and carbon dots as core via the sol-gel method for detection of ketoprofein indicated good target detection, appropriate PL characteristics and advantageous precise sensing [9–30].

3 Energy

3.1 Applications in Photovoltaic Solar Cells

Solar cells have less maintenance and operating cost compared to other technologies but require absorption in the range of the solar spectrum. The sensitized solar cells adsorb dye molecules on a porous semiconducting thin film imitating photosynthesis. QDs are of interest in SCs due to characteristics such as size, energy tunability, high-intensity light absorption/emission, phonon bottleneck effect, long relaxation time and distinct energy levels. The QD sensitized solar cells (QDSCs) features include ease of tunable absorption range as well as light, moisture, thermal stability [31, 38, 41, 45–53, 57–80, 106].

Different types of QDSCS include TiO_2/CdS, TiO_2/CdSe, TiO_2/CdS/CdSe/ZnS, TiO_2/CdS/CdSe:Cu/ZnS, TiO_2/CdS/CdSe:Mn/ZnS, TiO_2 NWs/$CuInS_2$, ZnO/Sb, ZnO/Sb_2S_3, Cu-doped, $CuInS_2$/TiO_2 nanodentrite, ZnO/Sb_2S_3, ZnTi/CdS, ZnONR@ TiO_2, TiO_2/CdS/CdSe/ZnS, TiO_2/ZnS/3CdSe/ZnSe, ZnSn/CdSe. The open circuit voltages (VOC) values range from 0.27 to 0.62 (V) and ŋ (%) from 1.52 to 8.18. A high efficiency of 5.6%, large current density versus voltage was obtained with large concentrations of Pbs QDs [31, 38, 41, 45–53, 67–80, 106].

3.2 *Applications in Photocatalysis*

There is growing interest in photocatalysis (PC) stimulated by hydrogen PC production which is in general via conventional steam reforming [89–93, 126].

QD applications in photocatalysis indicates that the N-doped graphene oxide quantum dots (combined with graphitic carbon nitride, doped with $BiVO_4$, TiO_2 as well as zinc sulfide QDs doped with different transition metals, using visible-light irradiation, UV-radiation 400 W) yields in general high stability for photocatalytic reactions, high photocatalytic activity for rhodamine B dye degradation, reusability without photoactivity loss, highest photocatalytic activity of 91.5%, photodegradation efficiency improved from 40 to 85%, 15.2% flux recovery ratio, improved durability and double dye removal [89–93, 126].

3.3 *QDs in Fuel Cells*

Various fuel cells use costly Pt catalysts tolerating only high purity hydrogen. However, the operating temperatures could be modified via catalysts, impure hydrogen, hydrocarbon-based fuels whereas the unique QDs can contribute to this technology allowing cheaper catalysts at anodes, cathodes [1, 7, 33, 59, 97–104].

Carbon Qds sensitized Pt@Bi_2WP_6/(fluorine-doped tin oxide) electrodes were used at the anode in direct methanol fuel cells via photoelectron-catalysts (visible-light-driven) for CH_3OH oxidation whereas electrodes indicated increase in photocurrent density. The CQD conductivity and stability were enhanced indicating increased tolerance towards CO species, less poisoning and boosted efficiency. Au-GQDs@Pt electrodes were used for CH_3OH oxidation at the anode indicating low charge transfer resistance, chemical/physical stability, electron conductivity, improved activity, stability of the catalysts, enhanced charge transfer from fuel to catalyst and platinum catalyst surface area [59, 97–104].

The functionalization of GQDs with chitosan, i.e. GQDs-CS, using alkaline solution for CH_3OH oxidation direct electrodeposition at the anode indicated excellent electrocatalytic activity for methanol redox reaction as well as an excellent improvement of redox reaction of methanol electrocatalytic activity [59, 97–104].

Using as an electrode Fe@Fe_2O_3/NCQDs at the anode via direct electrodeposition it was observed at steady state the maximum power generation 87% higher than nitrogen doped CQDs. At the cathode, sulfur and nitrogen co-doped GQDs indicate for the catalytic electron transfer of redox reaction an impressive enhancement of catalytic electron transfer as well as improved electrochemical performance. At the cathode, N-GQDs decorated on 3D MoS_2-rGO indicated enhanced electrocatalytic performance and superior CH_3OH tolerance properties as well as improved ORR stability. Also at the cathode using CQDs (CQDs@NS) as a catalyst for Cu foam indicated high power generation and excellent long-term stability [59, 97–104].

3.4 *QDs in Lithium-Based Batteries*

In the batteries charge transfer and appropriate electrode materials are essential for high efficiency/performance and in the lithium-based batteries QDs have introduced innovations whereas it is important to develop materials which have lower cost and indicate to be effective in electrochemical applications including batteries [81–88, 121–124].

The lithium-based batteries (Nb_2O_5@NC,TiO_2@GNRs) QDs, GQDs/HeCuO, $SnSe_2$QDs/rGO, (MoS_2@SnS-QDs/CNN) and the sodium-based batteries (Cu-MOC/NCNTs, TiO_2@GNRs) have excellent specific, reversible capacitance at the anode. The MoP QD@HCF lithium based and the NVOPF@C sodium-ion batteries also indicated good reversible capacity at the cathode [81–88, 121–124].

3.5 *Luminescent Perovskite Quantum Dots (LPQDs)*

Luminescent Perovskite Quantum Dots (LPQDs) are of considerable interest for areas including LEDS, lasers, illumination, displays, color-tunable white light production, photovoltaic cells, quantum computing due to their efficient/simple method of synthesis, tunable bandgap, high PLQY, optical purity, high spectral stability, blue-shifted emission peak, bioimaging, light emitting capability, tunable emission and narrower emission wavelengths [48, 60, 105].

3.6 Quantum Dot Device/Materials Discovery/Optimization Using AI

Optimizing device performance and identifying novel materials can be obtained from AI driven approaches by leveraging large datasets with information on QD structures, compositions, properties whereas analysis of the dataset's models can yield hidden correlations/patterns that serve as guides towards device configurations/promising materials predicting new QD material properties based only on the existing data. Among tools, machine learning is important to discover new nanomaterial features that can suggest novel compositions/structures likely to exhibit desired properties speeding up discovery with less time-consuming/costly experimental synthesis/characterization yielding improved operating conditions and more reliable/efficient devices [60].

3.7 QDs and Energy Storage

QDs participate in processes including lithium-sulfur batteries, supercapacitors and hydrogen production. In batteries of Li-S the low cost, high theoretical capacity/energy density meets requirements for high electric/portable/storage devices and can be used as positive electrodes/separators. However, practical applications suffer from challenges of discharge products, low conductivity, intermediate products, sulfur expansion. QDs have many surface functional sites/high specific surface area providing many active sites of localization/absorption of polysulfides accommodating volume expansion of sulfur particles and high sulfur load [54–62, 81–88, 121–126].

Supercapacitors (SCs) can offer long cycle life, fast charge/discharge, high power density and can in principle be divided into pseudo capacitors (hydrogen storage by redox reaction) and adsorption-desorption energy storage in double layer capacitors (DLCs). Research has been done to improve energy density (lower than traditional batteries) developing electrode materials with high capacity/strong cycle stability. Notable CQDs/QQDs retain stable chemical properties of carbon materials indicating quantum tunneling, surface/size effect, strong adsorption capacity for electrolyte ions and promising prospects in SCs [54–62, 81–88, 121–126].

The DLCs (charge storage), in general high ratio surface using carbon materials, whereas QDs can be applied in double-layer capacitors by embedding into materials (carbon electrodes) increasing thus carbon materials surface area in order to improve performance of SC. In addition, highly conductive crystalline, nanosized GQDs can shorten ion transmission path and when used to modify carbon nanofibers can produce great performance of electrochemical/mechanical performance of flexible supercapacitors adjusting crosslinking degrees between carbon nanofibers. Some research use QDs directly as electrodes with great advantage in micro SCs (MSCs) whereas N-GQDs, MoS_2-QDs acting as anode, cathode indicated very good electrochemical performance. Notable, GQDs can also be used as electrolytes in SCs [54–62, 81–88, 121–126].

The conventional catalysts low quantum efficiency affects applications at a large-scale of water decomposition via photocatalysis. In contrast, QDs improves substantially, photocatalytic activity compared to bulk semiconductor materials which can be attributed to multiple generation of excitons (from quantum confinement), improved, charge separation and transmission, unique physical, electronic, optical and excellent environmental stability properties.

P and B co-doped Si QDs indicated important quantum size effect on hydrogen production from photocatalysis. ZCS QDs also indicates excellent catalytic efficiency. Although C_3N_4 and CdS are great candidates (energy position, narrow band gaps) to produce H_2, due to fast recombination of holes and electrons from photogeneration, the hydrogen protection is low, whereas QDs could be used to increase the charge separation efficiency and improve the performance. Notable, QD surface modification and usage to capture carriers as co-catalysts can be used as catalytic reaction active site, reduce hydrogen evolution energy and improve charge separation efficiency [54–62, 81–88, 121–126].

4 QDs in Quantum Computing

QDs are very promising in quantum computing, serving as quantum information units which can exist simultaneously in multiple states due to the superposition property, allowing quantum computers to do, exponentially faster than classical computers, very complex calculations. An accurate control can be obtained of the spin/electronic states due to quantum confinement effects whereas the QDs can be manipulated to represent 1 and 0 states of classical bits, i.e. quantum computation basis [60–63, 107–120]. *Notable, AI, which is expected to be very important technology of the future, will require high computational capacity.*

4.1 Qubits from QDs in Quantum Computing

Notable, compared to other candidates for qubits, QDs are much more promising for quantum computing due to their maintenance of coherence (quantum state persistence over time) essential for quantum computing. Densely packed arrays of qubits can be obtained, due to their small size, making possible quantum processors construction, with high computational capacity necessary for quantum computing. The generation of QDs for qubits in QC requires methods such as colloidal synthesis. Notable, precise control of surface chemistry, size distribution and composition of QDs are essential in order to maintain quantum properties (efficient readout and qubit coupling) when integrating QDs and QC architectures eliminating decoherence/unwanted noise in qubit operations [60–63, 107–120].

4.2 Photonic Memory Using Quantum Dots in Unconventional Computing

In unconventional computing, photonic memory has been used in the von Neumann bottleneck whereas flash memory using perovskite QDs enabled electrical erasing of memory and photonic flash memory was attained emulating synaptic functions (long-term, short-term, spike-rate dependent plasticity, wavelength dependent synaptic weights) with great potential for perovskite-based quantum dot memory technologies and quantum information processing (spin-based quantum technologies) [60–63, 107–120].

4.3 Quantum Dots in Quantum Communication

In a secure manner, using quantum mechanics principles, quantum communication encodes/transmits information inherently safe from eavesdropping. QDs receivers, emitters permit quantum communication systems which are resistant to tampering or interception whereas single photons are detected/emitted as information can be encoded in individual photon quantum states transmitted between detector (receiver) and emitter (sender). During transmission the QDs quantum emitters must produce high purity/efficient single photons with preserved quantum communication information during transmission. Notably, perovskite QDs can emit precise energy level photons whereas size/position of QD can result in quantum operation errors [60–63, 107–120].

4.4 Quantum Gates from QD Arrays

In QC the quantum gates involves creating QD arrays (arranged and controlled) for quantum computations, and are segments which are essential in quantum circuits to perform these calculations. Similar to classical computing logical gates, the quantum gates are the quantum circuits fundamental building blocks which manipulate qubits quantum states, i.e. the basic units of quantum information used to do quantum operations. Whereas classical gates operate on (0 1), i.e. classical bits the quantum gates operate on qubits (in superposition). This allows simultaneous multiple states processing. As an example, the quantum NOT gate (X gate) flips the qubit state, from $|1>$ to $|0>$ and vice versa. The CNOT gate (control-NOT operations) and the Hadamard gate (creates superpositions) whereas complex operations can be done by combining in a quantum circuit multiple quantum gates based on reliable, accurate quantum gates. Notable to assure consistency, uniformity in quantum properties/calculations control of size, composition, shape of QDs is essential [60–63, 107–120].

4.5 *Quantum Algorithms from QDs*

The quantum algorithms based on QDs, which can represent/manipulate efficiently quantum states, introduces advanced methods to solve difficult problems not accessible to classical computations due to abilities such as algorithms (entanglement, superposition) which can enable parallel operations with exponential speedup and thus advantageously used in selected computing algorithms [60–63, 107–120].

4.6 *Quantum Memory Using QDs*

The QDs luminescent properties makes them appropriate for quantum memory applications whereas their efficient emission and absorption of photons allows them to manipulate/store quantum information for further processing/readout in quantum communication. Quantum memory is obtained by storing photons with encoded quantum information whereas dependent on the QDs coherence time the quantum information can be encoded, stored in the QDs, subsequently and read by suitable detectors making them appropriate to be quantum memory elements in quantum communication systems [60–63, 107–120].

References

1. Taft, C.A., de Lazaro, S.R. (eds.): Progress in Hydrogen Energy, Fuel Cells, Nano-Biotechnology and Advanced, Bioactive Compounds. Springer Nature, Switzerland AG, Cham (2024). https://doi.org/10.1007/978-3-031-75984_2
2. Taft, C.A., de Almeida, F.A. (eds.): Trends and Innovations in Energetic Sources, Functional Compounds and Biotechnology, Science, Simulation, Experiments. Springer Nature, Switzerland AG, Cham (2024). https://doi.org/10.1007/978-3-031-46545-1_2
3. Taft, C.A., de Lazaro, S.R. (eds.): Research Topics in Bioactivity, Environment and Energy, Experimental and Theoretical Tools. Springer Nature, Switzerland AG, Cham (2022). https://doi.org/10.1007/978-3-031-07622-0_2
4. La Porta, F.A., Taft, C.A. (eds.): Functional Properties of Advanced Engineering Materials and Biomolecules. Engineering Materials. Springer, Cham (2021). https://doi.org/10.1007/978-3-030-62226-8_2
5. La Porta, F.A., Taft, C.A. (eds.): Emerging Research in Science and Engineering Based on Advanced Experimental and Computational Strategies. Engineering Materials. Springer, Cham (2020). https://doi.org/10.1007/978-3-030-31403-3
6. Silva, J.C.F., dos Santos, J.D., Taft, C.A., Martins, J.B.L., Longo, E.: Stability of rolled-up GaAs nanotubes. J. Mol. Model. **23**, 204 (2017)
7. Canchaya, J.G.S., Furtado, N.C., Taft, C.A.: An overview of fuel cells and simulation models. Curr. Phys. Chem. **5**, 223–252 (2015)
8. Canchaya, J.G.S., Rebaza, A.V.G., Lemelle, D.S., Taft, C.A.: Theoretical studies of doped solid oxides for fuel cell applications. Curr. Phys. Chem. **4**, 45–45 (2014)
9. Yin, X., Liu, Z., Huanood, G., Sawatari, H., Shimamori, K., Kuragano, M., Tokuraku, K.: Analyzing amylin aggregation inhibition through quantum dot fluorescence imaging. Int. J. Mol. Sci. **25**, 11132 (2024). https://doi.org/10.3390/ijms252011132

10. Amiri, Z., Tamori, P., Alavi, K., Ghahramani, P., Cho, W.C., Farani, M.R., Huh, Y.S.: Quantum dot infused nanocomposites: revolutionizing diagnostic sensitivity. Nanoscale. **17**, 18477–18504 (2025). https://doi.org/10.1039/D5NR00440C
11. Tian, S., Hu, Z., Zhang, Z.: Modernization of charcoal drugs: integrating research paradigms of carbon dots to gain new perspectives. Chin. Herb. Med. **17**(2), 292–295 (2025). https://doi.org/10.1016/j.chmed.2025.01.004
12. Shi, L., Huanood, G., Miura, S., Kuragano, M., Tokuraku, K.: Real-time 3D imaging and inhibition analysis of human serum amyloid A aggregations using quantum dots. Int. J. Mol. Sci. **25**, 11128 (2024). https://doi.org/10.3390/ijms252011128
13. Wang, X., Wang, P., Li, M., Jian Li, J.: Advances in the preparation and biological applications of core@shell nanocrystals based on quantum dots and noble metal. RSC Adv. **14**, 26308–26324 (2024)
14. Liang, J., Yang, L., Ye, Z., Tang, Y., Ouyang, H., Yang, G., Lei, H., Lei, X.: CdSe/ZnS quantum dots-doped polymer optical fiber microprobe for pH sensing. Optic. Express. **33**, 2475 (2025)
15. Abdellatif, A.A.H., Younis, M.A.Y., Alsharidah, M., Rugaie, O.A., Tawfeek, H.M.: Biomedical applications of quantum dots: overview, challenges, and clinical potential. Int. J. Nanomed. **17**, 1951–1970 (2022)
16. Das, S., Mondai, S., Ghosh, D.: Carbon quantum dots in bioimaging and biomedicines. Front. Bioeng. Biotechnol. **11**, 1333752 (2024). https://doi.org/10.3389/fbioe.2023.1333752
17. Alam, N., Ihsan, H., Khan, S., Ullah, K.: Graphene quantum dots-based composites for biomedical applications, Chapter 4. In: Innovations and Applications of Hybrid Nanomaterials, pp. 62–90. IGI Global, New York (2024). https://doi.org/10.4018/979-8-3693-3268-9.ch004
18. Li, M., Huang, Y., Shen, C., Wang, Y., Lin, Y., Wang, Z., Chen, N., Luo, Y.: Application of quantum dots in cancer diagnosis and treatment: advances and perspectives. Nano Res. **18**, 94907163 (2025). https://doi.org/10.26599/NR.2025.94907163
19. Alhasan, A., Sani, S.F.A., TaJudidin, H.A., Ali, T.H.: Development of nanoparticles and quantum dots as alternative to iodinated contrast agents for mono-modal and bi-modal computed tomography imaging, nanomedicine: nanotechnology. Biol. Med. **68**, 102831 (2025)
20. Yang, H.L., Bai, L.-F., Geng, Z.-R., Chen, H., Xu, L.-T., Xie, Y.-C., Wang, D.-J., Gu, H.-W., Wang, X.M.: Carbon quantum dots: preparation, optical properties, and biomedical applications. Mater. Today Adv. **18**, 100376 (2023)
21. Salvi, A., Kharbanda, S., Thakur, P., Shandilya, M., Thakur, A.: Biomedical application of carbon quantum dots: a review. Carbon Trends. **17**, 100407 (2024)
22. Sk, M.P., Packirisamy, G., Misra, K., Hussain, S., Tariq, M.: Quantum dots for biological applications, volume II. Front. Bioeng. Biotechnol. https://doi.org/10.3389/978-2-8325-4668-0 (2024)
23. Thokchom, B., Bhavi, S.M., Abbigeri, M.B., Shettar, A.K., Yarajarla, R.B.: Green synthesis, characterization and biomedical applications of Centella asiatica-derived carbon dots. Carbon Lett. **33**, 1057–1071 (2023). https://doi.org/10.1007/s42823-023-00505-3
24. Zmejkoski, D.Z., Zdravkovic, N.M., Mitic, D.D., Markovic, Z.M., Budimir Filimonovic, M.D., Milivojevic, D.D., Todorovic Markovic, B.M.: Graphene quantum dots in bacterial cellulose hydrogels for visible light-activated antibiofilm and angiogenesis in infection management. Int. J. Mol. Sci. **26**, 1053 (2025). https://doi.org/10.3390/ijms26031053
25. Bazmi, A., Hamsian-Etefaq, R., Hashemi-Nasab, S., Asefnejad, A., Sadeq, A.M., Ashkani, O.: Role of quantum dots in the advancement of biomedical engineering: a general review. Adv. J. Chem. Sect. A. **8**(8), 1385–1397 (2025)
26. Nasab, S.Z., Rahi, A., Ghaee, A., Koohkhezri, M., Farani, M.R., Gholipour, H.M., Atabaki, A.H., Hamblin, M.R., Mostafavi, E., Kang, H.: Antimicrobial carbon materials-based quantum dots: from synthesis strategies to antibacterial properties for diagnostic and therapeutic applications in wound healing. Coord. Chem. Rev. **522**, 216211 (2025)
27. Alluhaidan, T., Garcia, I.M., Alexis, M., Qaw, M., Collares, F.M., Williams, M.A., Melo, M.A.S.: Unlocking tiny titans: 360 view of the quantum dots nanotechnology for dental applications. Front. Dent. Med. **6**, 1503057 (2025). https://doi.org/10.3389/fdmed.2025.1503057

28. Hu, Y., Wang, X., Niu, Y., He, K., Tang, M.: Application of quantum dots in brain diseases and their neurotoxic mechanism. Nanoscale Adv. **6**, 3733 (2024)
29. Chen, J., Yin, S., Yang, F., Guo, S., Zhang, J., Lu, Z., Gao, T.: A single-molecule graphene quantum dot: a novel efficient photosensitizer for photodynamic cancer therapy. Chem. Sci. **16**(30), 13923–13934 (2025). https://doi.org/10.1039/S5SC03226A
30. Huong, L.T.T., Hung, N.P., Ha, N.T., Luyen, N.T., Hien, N.T., Ca, N.X., Thuy, N.T.M.: Chemically synthesized CdSe quantum dots induce apoptosis in AGS gastric cancer cells via ROS generation. Nanoscale Adv. **7**, 572–582 (2025)
31. Zhonglin, D., Ma, D.: Recent progress in I-III-VI colloidal quantum dots integrated solar cells. Curr. Opin. Colloid Interface Sci. **75**, 101890 (2025)
32. Chakrabarti, A., Majumdar, P., Alessandri, E., Roemer, C.: Recent advancement of quantum dot-based nanocomposites as electrode materials for secondary batteries. Energies. **18**, 630 (2025). https://doi.org/10.3390/en18030630
33. Pradhan, A.K., Halder, S., Chakraborty, C.: Pt-nanoparticles on ZnO/carbon quantum dots: a trifunctional nanocomposite with superior electrocatalytic activity boosting direct methanol fuel cells and zinc–air batteries. J. Mater. Chem. A. **13**, 243 (2025)
34. Ubbink, R.F., Speelman, T., Esteban, D.N., van Leeuwen, M., Stam, M., Bals, S.A., De Wijs, G., van Eck, E.R.H., Houtepen, A.J.: Phosphorus oxidation controls epitaxial shell growth in InP/ZnSe quantum dots. ACS Nano. **19**, 1150–1158 (2025)
35. Naylor-Aldamson, L., Price, T.W., Booth, Z., Leonard, S., Gallo, J., Tung, L.D., et al.: PEGylation of indium phosphide quantum dots prevents quantum dot mediated platelet activation. J. Mater. Chem. B. **13**, 1052 (2025)
36. Santos, T.F., Souza, D.F.S., Santos, E.V., Carvalho, B.R., Nascimento, J.H.O.: Graphene and graphene quantum dots applied to batteries and supercapacitors. Nano Trends. **9**, 100077 (2025)
37. Ye, T., Tian, D., Dan Wu, D., Wei Sun, X., Wang, K.: Submicron quantum dot light-emitting diodes enabled by pixelated topological meta-mirror. Nanophotonics. **14**, 221 (2025). https://doi.org/10.1515/nanoph-2024-0543
38. Kärcher, V., Reiker, T., da Costa, P.F.G.M., de Camargo, A.S.S., Zacharias, H.: Quantum control in size selected semiconductor quantum dot thin films. Nanophotonics. **14**(2), 229–239 (2025). https://doi.org/10.1515/nanoph-2024-0529
39. Rathee, G., Puertas-Segura, A., Blair, J., Rathee, J., Tzanov, T.: Quantum dots@layered double hydroxides: emerging nanocomposites for multifaceted applications. Prog. Mater. Sci. **150**, 101403 (2025)
40. Peng, Y., Gu, W., Dong, Y., Ji, Y., Yan, D., Lin, Z., Zhang, L., Zou, Y., Zeng, H.: 15.13% Sr-doped CsPbI3 quantum dots with near-unity quantum yield via surface ligand compensation. NanoResearch. **18**, 94907257 (2025). https://doi.org/10.26599/NR.2025.94907257
41. Liu, J., Zhao, D., Wu, X., Wu, D., Su, N., Wang, F., Chen, F., Fu, C., Wang, J., Zhang, Q.: Synergistic dual-defect band engineering for highly efficient photocatalytic degradation of microplastics via Nb-induced oxygen vacancies in SnO2 quantum dots. J. Mater. Chem. A. **13**(3), 4429 (2025). https://doi.org/10.1039/d4ta07579j
42. Appel, M.H., Ghorbal, A., Shofer, N., Zaporski, L., Manna, S., da Silva, S.F.C., Haeusler, U., Le Gall, C., Rastelli, A., Gangloff, D.A., Atatüre, M.: A many-body quantum register for a spin qubit. Nat. Phys. **21**, 368 (2025). https://doi.org/10.1038/s41567-024-02746-z
43. Kapoor, S., Rodek, A., Mikołajczyk, M., Szuniewicz, J., Filip Sośnicki, F., Kazimierczuk, T., Piotr Kossacki, P., Karpiński, M.: Electro-optic frequency shift of single photons from a quantum dot. Nanophotonics. **14**(11), 1775–1782 (2025). https://doi.org/10.1515/nanoph-2024-0550
44. Rickert, L., Żołnacz, K., Vajner, D.A., von Helversen, M., Rodt, S., Reitzenstein, S., Liu, H., Li, S., Ni, H., Wyborski, P., Sęk, G., Musiał, A., Niu, Z., Heindel, T.: A fiber-pigtailed quantum dot device generating indistinguishable photons at GHz clock-rates. Nanophotonics. **14**(11), 1795–1808 (2025). https://doi.org/10.1515/nanoph-2024-0519

45. Ghasemzadeh, R., Akhbari, K.: Degradation of acid blue 41 with carbon quantum dots@ MOF-808 nanocomposite as a biocompatible photocatalyst under visible light. J. Photochem. Photobiol. A Chem. **458**, 115984 (2025)
46. Yu, J., Hilmi Volkan Demir, H.V., Sharma, M.: Optical signatures of lattice strain in chemically doped colloidal quantum wells. Nat. Commun. **16**, 823 (2025). https://doi.org/10.1038/s41467-025-55984-x
47. Sharma, S.N., Semalti, P., Bhawna, A., Rao, S.: Pioneering advancements in quantum dot solar cells: innovations in synthesis and cutting-edge applications. Curr. Opin. Colloid Interface Sci. **77**, 101905 (2025)
48. Mirea, A.G., Vlaicu, I.D., Derbali, S., Neatu, F., Tomulescu, A.G., Besleaga, C., Enculescu, M., Kuncser, A.C., Iacoban, A.C., Filipoiu, N., Cuzminschi, M., Nemnes, G.A., Manolescu, A., Florea, M., Pintilie, L.: Electron transporting bilayers for perovskite solar cells: spray coating deposition of c-TiO2/m-SnO2-quantum dots. Colloids Surf. A Physicochem. Eng. Asp. **705**, 135508 (2025)
49. Maitra, S.K., Raju, K., Swathi, G.V., Thrinath, B.V.S., Choudhury, U., Budidha, N., Aegeegn, D.B.: A theoretical exploration of quantum dots and nanowires as next-generation photovoltaics. Solar Energy Adv. **5**, 100085 (2025)
50. El El Ghazi, H., En-nadir, G.R.: Quantum dots: the new frontiers in science and technology. Ergon Int. J. (2024). https://doi.org/10.23880/eoij-16000324. https://www.researchgate.net/publication/379961047
51. Huang, T., He, S., Ni, A., Lian, T., Tang, M.L.: Triplet energy transfer from quantum dots increases Ln(III) photoluminescence, enabling excitation at visible wavelengths. Chem. Sci. **15**, 455 (2024)
52. Zhang, Z., Wang, W., Rao, H., Pan, Z., Zhong, X.: Improving the efficiency of quantum dot-sensitized solar cells by increasing the QD loading amount. Chem. Sci. **15**, 5482 (2024)
53. Mawaddah, F.A.N., Bisri, S.Z.: Advancing silver bismuth sulfide quantum dots for practical solar cell applications. Nanomaterials. **14**, 1328 (2024). https://doi.org/10.3390/nano14161328
54. Galář, P., Kopenec, J., Král, R., Matejka, F., Zemenová, P., Dopita, M., Hapala, P., König, D., Vrbka, P., Kůsová, K.: Engineering the thermal and energy-storage properties in quantum dots using dominant faceting: the case study of silicon. ACS Nano. **19**, 2196–2212 (2025)
55. Guo, R., Yuan, X., Zhou, X., Chen, H., Xie, H., Hu, Q., Luo, H., Zhang, D.: All-organic quantum dots-boosted energy storage density in PVDF-based nanocomposites via dielectric enhancement and loss reduction. Polymers. **17**, 390 (2025). https://doi.org/10.3390/polym17030390
56. Abdelsalam, H., Zhang, Q.F.: Properties and applications of quantum dots derived from two-dimensional materials. Adv. Phys.: X. **7**, 2048966 (2022). https://doi.org/10.1080/23746149.2022.2048966
57. Zhu, D., Jiang, S., Wang, Y., Liu, D., Bao, W., Liu, L., Qu, J., Wang, Y., Liao, C.: Three-dimensional direct lithography of stable quantum dots in hybrid glass. Int. J. Extrem. Manuf. **7**, 035503 (2025). https://doi.org/10.1088/2631-7990/adaab1
58. Bhujbal, P.K., Abhijit, T., Supekar, A.T., Kadam, P.A., Vashishth, N., Mujawar, A., Singh, U., Ray, B., Mahadik, S.A., Datar, S., Majumdar, B., Patole, S.P., Dhirhe, D., Pathan, H.M.: Enhancing charge transfer in hybrid solar cells: the role of pulse laser-assisted hydrothermally synthesized Au@N-S-doped fluorescent carbon quantum dots as Forster resonance energy transfer antennas. Adv. Compos. Hybrid Mater. **8**, 168 (2025). https://doi.org/10.1007/s42114-025-01256-7
59. Al Murisi, M., Tawalbeh, M., Al-Saadi, R., Yasin, Z., Temsah, O., Al-Othman, A., Rezakazemi, M., Olabi, A.G.: New insights on applications of quantum dots in fuel cell and electrochemical systems. Int. J. Hydrog. Energy. **52**, 52694E73 (2024). https://doi.org/10.1016/j.ijhydene.2023.03.020
60. Jayan, D., Babu, K.: Luminescent perovskite quantum dots: progress in fabrication, modelling and machine learning approaches for advanced photonic and quantum computing applications. J. Lumin. **277**, 120906 (2025)

61. Khadilkar, P., Samudre, N.S., Krishnamurty, S.: Quasi-molecular hydrogen storage capacity of graphene quantum dots: a dispersion corrected DFT study. J. Energy Storage. **84**, 110833 (2024)
62. Xu, Q., Niu, Y., Li, J., Yang, Z., Gao, J., Ding, L., Ni, H., Zhu, P., Liu, Y., Tang, Y., Lv, Z.-P., Peng, B., Hu, T.S., Zhou, H., Xu, C.: Recent progress of quantum dots for energy storage applications. Carbon Neutr. **1**, 13 (2022). https://doi.org/10.1007/s43979-022-00002-y
63. Chan, M.L., Bell, T.J., Pettersson, L.A., Chen, S.X., Yard, P., Sørensen, A.S., Paesani, S.: Tailoring fusion-based photonic quantum computing schemes to quantum emitters. PRX Quant. **6**, 020304 (2025). https://doi.org/10.1103/PRXQuantum.6.020304
64. Chen, Y.C., Chiang, W.H., Kurniawan, D., Yeh, P.C., Otake, K., Kung, C.W.: Impregnation of graphene quantum dots into a metal-organic framework to render increased electrical conductivity and activity for electrochemical sensing. ACS Appl. Mater. Interfaces. **11**(38), 35319–35326 (2019). https://doi.org/10.1021/acsami.9b11447
65. Niu, Y., Wang, J., Zhang, J., Shi, Z.: Graphene quantum dots as a novel conductive additive to improve the capacitive performance for supercapacitors. J. Electroanal. Chem. **828**, 1–10 (2018). https://doi.org/10.1016/j.jelechem.2018.09.003
66. Campuzano, S., Yanez-Sede, P., Pingarron, J.M.: Carbon dots and graphene quantum dots in electrochemical biosensing. Nanomaterials. **9**, 634 (2019). https://doi.org/10.3390/nano9040634
67. Phuc, D.H., Tung, H.T.: Quantum dot sensitized solar cell based on the different photoelectrodes for the enhanced performance. Solar Energy Mater. Solar Cells. **196**, 78–83 (2019). https://doi.org/10.1016/j.solmat.2019.03.038
68. Wang, M., et al.: Current density enhancement for quantum dot-sensitized solar cells by modulation on the quantum dot loading amount of anatase nanowire array photoelectrodes. J. Solid State Electrochem. **25**(7), 2087–2096 (2021). https://doi.org/10.1007/s10008-021-04969-z
69. Sun, Z., et al.: Band energy modulation on Cu-doped Sb2S3-based photoelectrodes for charge generation and transfer property of quantum dotesensitized solar cells. J. Nanopart. Res. **22**(9), 282 (2020). https://doi.org/10.1016/j.mssp.2021.106219
70. Wang, M., et al.: Improving loading amount of nanodendrite array photo-electrodes on quantum dot sensitized solar cells by second electrochemical deposition. Mater. Sci. Semicond. Process. **137**, 106219 (2022). https://doi.org/10.1016/j.mssp.2021.106219
71. Peng, Z., et al.: Enhancement on charge generation and transfer properties of Sb2S3 quantum-dot-sensitized solar cells by ZnO nanorods optimization. J. Electron. Mater. **50**(1), 100–107 (2021). https://doi.org/10.1007/s11664-020-08542-8
72. Khodam, F., Amani-Ghadim, A.R., Aber, S.: Preparation of CdS quantum dot sensitized solar cell based on ZnTi-layered double hydroxide photoanode to enhance photovoltaic properties. Solar Energy. **181**, 325–332 (2019). https://doi.org/10.1016/j.solener.2019.02.019
73. Liu, Y., et al.: Highly efficient quantum-dot-sensitized solar cells with composite semiconductor of ZnO nanorod and oxide inverse opal in photoanode. Electrochim. Acta. **412**, 140145 (2022). https://doi.org/10.1016/j.electacta.140145
74. Li, W., Zhang, S., Chen, Q., Zhong, Q.: Highly-dispersed CoS2/N-doped carbon nanoparticles anchored on RGO skeleton as a hierarchical composite counter electrode for quantum dot sensitized solar cells. Chem. Eng. J. **430**, 132732 (2022). https://doi.org/10.1016/j.cej.2021.132732
75. Bin Jin, B., et al.: Antimony tin oxide/lead selenide composite as efficient counter electrode material for quantum dot-sensitized solar cells. J. Colloid Interface Sci. **598**, 492–499 (2021). https://doi.org/10.1016/j.jcis.2021.04.073
76. Muthalif, M.P.A., Choe, Y.: Surface modification of CuS counter electrodes by hydrohalic acid treatment for improving interfacial charge transfer in quantum-dot-sensitized solar cells. J. Colloid Interface Sci. **595**, 15–24 (2021). https://doi.org/10.1016/j.jcis.2021.03.113
77. Bin Jin, B., et al.: Voltage-assisted SILAR deposition of CdSe quantum dots to construct a high performance of ZnS/CdSe/ZnS quantum dot-sensitized solar cells. J. Colloid Interface Sci. **586**, 640–646 (2021). https://doi.org/10.1016/j.jcis.2020.10.132

78. Cao, J., et al.: Preparation and photovoltaic properties of CdS quantum dot-sensitized solar cell based on zinc tin mixed metal oxides. J. Colloid Interface Sci. **498**, 223–228 (2017). https://doi.org/10.1016/j.jcis.2017.03.061
79. Suriyawong, N., Aragaw, B., Shi, J.-B., Lee, M.-W.: Ternary CuBiS2 nanoparticles as a sensitizer for quantum dot solar cells. J. Colloid Interface Sci. **473**, 60–65 (2016). https://doi.org/10.1016/j.jcis.2016.03.062
80. Mahalingam, S., et al.: Mixture deposition method for graphene quantum dots-based dye-sensitized solar cell. Electrochim. Acta. **404**, 139732 (2022). https://doi.org/10.1016/j.electacta.2021.139732
81. Lian, Y., et al.: Nb2O5 quantum dots coated with biomass carbon for ultra-stable lithium-ion supercapacitors. J. Alloys Compd. **850**, 156808 (2021). https://doi.org/10.1016/j.jallcom.2020.156808
82. Yu, W.-J., et al.: Confinement of TiO2 quantum dots in graphene nanoribbons for high-performance lithium and sodium ion batteries. J. Alloys Compd. **898**, 162856 (2022). https://doi.org/10.1016/j.jallcom.2021.162856
83. Kim, J., Jang, W., Kim, J.H., Yang, C.-M.: Synthesis of graphene quantum dots-coated hierarchical CuO microspheres composite for use as binder-free anode for lithium-ion batteries. Compos. B Eng. **222**, 109083 (2021). https://doi.org/10.1016/j.compositesb.2021.109083
84. Huang, J., et al.: Enhanced specific energy in fast-charging lithium-ion batteries negative electrodes via Ti-O covalency-mediated low potential. Nat. Commun. **16**(1), 6243 (2025)
85. Wei, M., et al.: A 3D free-standing thin film based on N, P-codoped hollow carbon fibers embedded with MoP quantum dots as high efficient oxygen electrode for Li-O2 batteries. Energy Storage Mater. **17**, 226–233 (2019). https://doi.org/10.1016/j.ensm.2018.07.012
86. Xiang Huang, Z., Liu, B., Kong, D., Wang, Y., Ying Yang, H.: SnSe2 quantum dot/rGO composite as high performing lithium anode. Energy Storage Mater. **10**, 92–101 (2018). https://doi.org/10.1016/j.ensm.2017.08.008
87. Deng, G., et al.: Graphene quantum dots-shielded Na3(VO) 2(PO4)2F@C nanocuboids as robust cathode for Na-ion battery. Energy Storage Mater. **5**, 198–204 (2016). https://doi.org/10.1016/j.ensm.2016.07.007
88. Ke, G., et al.: Ultrathin MoS2 anchored on 3D carbon skeleton containing SnS quantum dots as a high-performance anode for advanced lithium ion batteries. Chem. Eng. J. **403**, 126251 (2021). https://doi.org/10.1016/j.cej.2020.126251
89. Zhang, X., Li, Z., Xu, S., Ruan, Y.: Carbon quantum dot-sensitized hollow TiO2 spheres for high-performance visible light photocatalysis. New J. Chem. **45**(19), 8693–8700 (2021). https://doi.org/10.1039/D1NJ00501D
90. Zhou, Q., et al.: Novel hierarchical carbon quantum dots-decorated BiOCl nanosheet/carbonized eggshell membrane composites for improved removal of organic contaminants from water via synergistic adsorption and photocatalysis. Chem. Eng. J. **420**, 129582 (2021). https://doi.org/10.1016/j.cej.2021.129582
91. Liu, W., Ning, C., Sang, R., Hou, Q., Ni, Y.: Lignin-derived graphene quantum dots from phosphous acid-assisted hydrothermal pretreatment and their application in photocatalysis. Ind. Crop. Prod. **171**, 113963 (2021). https://doi.org/10.1016/j.indcrop.2021.113963
92. Shi, W., et al.: Onion-ring-like g-C3N4 modified with Bi3TaO7 quantum dots: a novel 0D/3D S-scheme heterojunction for enhanced photocatalytic hydrogen production under visible light irradiation. Renew. Energy. **182**, 958–968 (2022). https://doi.org/10.1016/j.renene.2021.11.030
93. Vatanpour, V., Karatas, O., Amiri, S., Rajabi, H.R., Koyuncu, I., Khataee, A.: Different metal-doped ZnS quantum dots photocatalysts for enhancing the permeability and antifouling performances of polysulfone membranes with and without UV irradiation. Chemosphere. **294**, 133705 (2022). https://doi.org/10.1016/j.chemosphere.2022.133705.pho
94. Yang, J., et al.: Synthesis of three-dimensional Au-graphene quantum dots@Pt core-shell dendritic nanoparticles for enhanced methanol electro-oxidation. Nanotechnology. **30**(49), 495706 (2019). https://doi.org/10.1088/1361-6528/ab3dc3

95. Farahmand Habibi, M., Arvand, M., Sohrabnezhad, S.: Boosting bioelectricity generation in microbial fuel cells using metal@metal oxides/nitrogen-doped carbon quantum dots. Energy. **223**, 120103 (2021). https://doi.org/10.1016/j.energy.2021.120103
96. Zheng, H., Niu, P., Zhao, Z.: Carbon quantum dot sensitized Pt@Bi 2 WO 6/FTO electrodes for enhanced photoelectrocatalytic activity of methanol oxidation. RSC Adv. **7**(43), 26943–26951 (2017). https://doi.org/10.1039/C7RA01867C
97. Kappen, J., Krishnan, G.N.S., Kumar, R., et al.: Graphene quantum dots-anchored Au-Pt bimetallic nanoparticle: a benign electrocatalyst for sustainable fuel cell applications. Colloids Surf. A Physicochem. Eng. Aspects. **725**, 137517 (2025)
98. Hasanzadeh, M., Shadjou, N., Marandi, M.: Graphene quantum dot functionalized by chitosan and betacyclodextrin as a new support nanocomposite material for efficient methanol electrooxidation. J. Alloys Compd. **688**, 171–186 (2016). https://doi.org/10.1016/j.jallcom.2016.07.202
99. Das, S.K., et al.: Morphological evolution of carbon quantum dots to carbon nanoneedles and N/F codoping as an efficient catalyst for oxygen electrochemistry. ACS Appl. Eng. Mater. **2**(7), 1894–1907 (2024)
100. Zhou, L., Fu, P., Wang, Y.: Microbe-engaged synthesis of carbon dot-decorated reduced graphene oxide as high performance oxygen reduction catalysts. J. Mater. Chem. **4**, 7222–7229 (2016). https://doi.org/10.1039/C6TA01662F
101. Zhang, P., Wei, J.S., et al.: Heteroatom-doped carbon dots based catalysts for oxygen reduction reactions. J. Colloid Interface Sci. **537**, 716 (2019). https://doi.org/10.1016/j.jcis.2018.11.024
102. Huang, Y., Tan, Z., Wu, H., Feng, C., Ding, Y.: Fe @ Pt core-shell nanoparticles as electrocatalyst for oxygen reduction reaction in acidic media. Ionics. **24**, 229–236 (2018). https://doi.org/10.1007/s11581-017-2186-7
103. Lan, D., et al.: N, S co-doped carbon quantum dots anchoring on copper-vacancy-rich Cu nanowires/Cu foam as the cathode in microbial fuel cells: role of C-S-Cu active site. Sci. Total Environ. **805**, 150340 (2022). https://doi.org/10.1016/j.scitotenv.2021.150340
104. Zhang, M., Song, Z., Wang, Z., Wang, A., Zhu, G., Shao, S.: Platinum quantum dots enhance electrocatalytic activity of bamboo-like nitrogen doped carbon nanotubes embedding Co-MnO nanoparticles for methanol/ethanol oxidation. J. Colloid Interface Sci. **590**, 164–174 (2021). https://doi.org/10.1016/j.jcis.2021.01.045
105. Sung, C.H., Huang, S.D., Kumar, G., Lin, W.C., Lin, C.C., Kuo, H.C., Chen, F.C.: Highly luminescent perovskite quantum dots for light-emitting devices: photopatternable perovskite quantum dot–polymer nanocomposites. J. Mater. Chem. C. **10**(42), 15941–15947 (2022)
106. Li, H., Zhang, W.: Perovskite tandem solar cells: from fundamentals to commercial deployment. Chem. Rev. **120**(18), 9835–9950 (2020)
107. Chen, M., Yin, Z., Shan, Z., Zheng, X., Liu, L., Dai, Z., Zhang, J., Liu, S.F., Xu, Z.: Application of machine learning in perovskite materials and devices: a review. J. Energy Chem. **94**, 254–272 (2024)
108. de Arquer, F.P.G., Talapin, D.V., Klimov, V.I., Arakawa, Y., Bayer, M., Sargent, E.H.: Semiconductor quantum dots: technological progress and future challenges. Science. **373**(6555), eaaz8541 (2021)
109. Kagan, C.R., Bassett, L.C., Murray, C.B., Thompson, S.M.: Colloidal quantum dots as platforms for quantum information science. Chem. Rev. **121**(5), 3186–3233 (2020)
110. Wang, Y., Lv, Z., Chen, J., Wang, Z., Zhou, Y., Zhou, L., Chen, X., Han, S.T.: Photonic synapses based on inorganic perovskite quantum dots for neuromorphic computing. Adv. Mater. **30**(38), 1802883 (2018)
111. Harankahage, D., Cassidy, J., Yang, M., Porotnikov, D., Williams, M., Kholmicheva, N., Zamkov, M.: Quantum computing with exciton qubits in colloidal semiconductor nanocrystals. J. Phys. Chem. C. **125**(40), 22195–22203 (2021)
112. Wang, Y., Ding, G., Mao, J.Y., Zhou, Y., Han, S.T.: Recent advances in synthesis and application of perovskite quantum dot based composites for photonics, electronics and sensors. Sci. Technol. Adv. Mater. **21**(1), 278–302 (2020)

113. Castelletto, S., De Angelis, F., Boretti, A.: Prospects and challenges of quantum emitters in perovskites nanocrystals. Appl. Mater. Today. **26**, 101401 (2022)
114. Almutlaq, J., Liu, Y., Mir, W.J., Sabatini, R.P., Englund, D., Bakr, O.M., Sargent, E.H.: Engineering colloidal semiconductor nanocrystals for quantum information processing. Nat. Nanotechnol. **19**, 1091–1100 (2024)
115. Liu, X., Tao, L., Mei, S., Cui, Z., Shen, D., Sheng, Z., Yu, J., Ye, P., Zhi, T., Tao, T., Wang, L.: White-light GaN-μLEDs employing green/red perovskite quantum dots as color converters for visible light communication. Nanomaterials. **12**(4), 627 (2022)
116. Gao, K., Li, Y., Yang, Y., Liu, Y., Liu, M., Liang, W., Zhang, B., Wang, L., Zhu, J., Wu, K.: Manipulating coherent exciton dynamics in CsPbI3 perovskite quantum dots using magnetic field. Adv. Mater. **36**(6), 2309420 (2024)
117. Stefanatos, D., Karanikolas, V., Paspalakis, E.: Modelling the optical control of electron spin dynamics in a quantum dot near a two-dimensional semiconductor. J. Phys. Conf. **1730**(1), 012046 (2021)
118. Yen, M.C., Lee, C.J., Liu, K.H., Peng, Y., Leng, J., Chang, T.H., Chang, C.C., Tamada, K., Lee, Y.J.: All-inorganic perovskite quantum dot light-emitting memories. Nat. Commun. **12**(1), 4460 (2021)
119. Liu, Z., Qin, X., Chen, Q., Jiang, T., Chen, Q., Liu, X.: Metal–halide perovskite nanocrystal superlattice: self-assembly and optical fingerprints. Adv. Mater. **35**(16), 2209279 (2023)
120. Li, Q., Zhang, Y., Yu, Y., Chen, Z., Jin, L., Li, Y., Li, T., Yang, Y., Zhao, H., Li, J., Dai, H.: Light enhanced low-voltage nonvolatile memory based on all-inorganic perovskite quantum dots. Nanotechnology. **30**(37), 37LT01 (2019)
121. Li, H., et al.: Ultra-fast transfer and high storage of Li+/Na+ in MnO quantum dots@carbon hetero-nanotubes: appropriate quantum dots to improve the rate. Energy Storage Mater. **17**, 157–166 (2019). https://doi.org/10.1016/j.ensm.2018.07.021
122. Bao, W., et al.: Vapor phase infiltration of ZnO quantum dots for all-solid-state PEO-based lithium batteries. Energy Storage Mater. **43**, 258–265 (2021). https://doi.org/10.1016/j.ensm.2021.09.010
123. Huang, S., Wang, M., Jia, P., Wang, B., Zhang, J., Zhao, Y.: N-graphene motivated SnO2@SnS2 heterostructure quantum dots for high performance lithium/sodium storage. Energy Storage Mater. **20**, 225–233 (2019). https://doi.org/10.1016/j.ensm.2018.11.024
124. Ma, F., et al.: Interface engineering of heterostructural quantum dots towards high-rate and long-life lithium-sulfur batteries. Nano Energy. **133**, 110445 (2025)
125. Yang, S., Liu, C., Li, H., Wang, S., Choi, J., Li, L.: Quantum dot heterostructure with directional charge transfer channels for high sodium storage. Energy Storage Mater. **39**, 278–286 (2021). https://doi.org/10.1016/j.ensm.2021.04.039
126. Wu, T., et al.: Application of QD-MOF composites for photocatalysis: energy production and environmental remediation. Coord. Chem. Rev. **403**, 213097 (2020). https://doi.org/10.1016/j.ccr.2019.213097

Physics of Some Selected Non-perovskite ABO_3 Compounds

Sudhindra Rayaprol and Shovit Bhattacharya

Abstract Non-perovskite ABO_3 compounds such as bixbyite $FeMnO_3$ or ilmenite $FeTiO_3$ are known to exhibit interesting structural and physical properties. Structure and physical properties are strongly correlated and also depends upon the composition of the compounds. These compounds have found interesting applications as anode materials for lithium-ion batteries, magnetoelectrics etc. Structural and physical these compounds are reviewed here to bring out their application potential and future scope in designing materials for next generation electronics, catalysts and magnetic devices.

1 Introduction

In simple words, Perovskite refers to the mineral calcium titanate, having the composition $CaTiO_3$, which was first discovered by Gustav Rose in 1893 from the deposits found in the Ural Mountains of Russia. The mineral $CaTiO_3$ has been thus named Perovskite as dedication to famous Russian mineralogist, Lev Alekseyevich von Perovski [1, 2]. The structure of perovskite compound was nicely explained by Victor Goldschmidt in his work on tolerance factors [3]. The general formula for perovskites is given as ABX_3, where A is large cation (usually an alkaline metal such

S. Rayaprol (✉)
UGC-DAE Consortium for Scientific Research—Mumbai Centre, BARC Campus, Trombay, Mumbai, India

Savitribai Phule Pune University, Ganeshkhind, Pune, India
e-mail: sudhindra@csr.res.in

S. Bhattacharya
Technical Physics Division, Bhabha Atomic Research Centre, Trombay, Mumbai, India

Homi Bhabha National Institute, Anushaktinagar, Mumbai, India
e-mail: shovitb@barc.gov.in

C. A. Taft, S. R. d. Lazaro (eds.), *Innovative Materials for Energy and Sustainability*, Engineering Materials,
https://doi.org/10.1007/978-3-032-09149-9_3

as Li, Na, K or alkaline earth metal such as Ca, Ba, Sr or lanthanide such as La, Nd, Ho etc.), *B* is a transition metal ion (such as Mn, Fe, Co, etc.) and X is an oxide or a halide ion (Cl, F etc.). Majority of perovskite compounds studied are either oxides or fluorides [4], but other halides such as chlorides, bromides, and other anions such as hydrides and sulphides are also known to exist [5]. We shall however restrict our discussions to few selected oxide compounds only in this chapter.

2 Perovskite Compounds and Charge Neutrality

The ionic compounds containing cations and anions must be electrically neutral. If q_A, q_B are charges on cations and q_O on the anion (oxide in this case), then the overall the charges can be balanced as: $q_A + q_B = -3q_O$. In simpler words, based on charge balancing in the ABO_3 compounds, *A* site can accommodate cations having 1^+, 2^+ or 3^+ charge states, while correspondingly *B* site can have cations with 5^+, 4^+ and 3^+ charge states, respectively. In addition to this, *B* site can also accommodate a 6^+ ion, such as Re or Mo. It is, of course, not certain that all the compositions in the same combination will form in the same structure type or symmetry. The final structure of the compound, as we shall see later in this chapter, depends upon various parameters, combination of cations/anions and their charges being one of them. Thus, different combinations of cations can be used to form structures as shown in Table 1.

The *A*-site cations are usually valence stable ions (1^+, 2^+ or 3^+), however the *B*-site cations sometimes are known to have more than one valence state, for example, Fe. Fe can exist in 2^+ or 3^+ and in some unusual cases the rare 4^+ state also (as in $SrFeO_3$). This shows that in case of perovskite oxides, the mixed valence states of the *B* transition metal ions can also be stabilized [6].

Table 1 Simple perovskite oxides classified based on charge combinations at *A* and *B* sites

Sl. no.	Charge combinations	Stoichiometry	Compounds	Crystal structure
1.	$A_1^{3+}, B_1^{3+}, O_3^{2-}$	ABO_3	$LaAlO_3$ $NdGaO_3$ $BiFeO_3$	Rhombohedral ($R\bar{3}c$) Orthorhombic (*Pnma*) Rhombohedral (*R3c*)
2.	$A_1^{2+}, B_1^{4+}, O_3^{2-}$	ABO_3	$SrTiO_3$ $BaTiO_3$ $CaTiO_3$	Cubic ($Pm\bar{3}m$) Tetragonal (*P4mm*) Orthorhombic (*Pnma*)
3.	$A_1^{1+}, B_1^{5+}, O_3^{2-}$	ABO_3	$KTaO_3$ $NaNbO_3$ $KNbO_3$	Cubic ($Pm\bar{3}m$) Cubic ($Pm\bar{3}m$) Cubic ($Pm\bar{3}m$)
4.	B_1^{6+}, O_3^{2-}	BO_3	ReO_3	Cubic ($Pm\bar{3}m$)

However, when A-site is partially substituted by another cation with a different valence state, then the B-site cation can experience a mixed valence state. Say for example in $LaMnO_3$ (=ABO_3) the usual valence can be written as: $A^{3+}B^{3+}O^{2-}{}_3$ i.e., $La^{3+}Mn^{3+}O^{2-}{}_3$. Now if in this composition any aliovalent substitution (Ca^{2+} or Sr^{2+}) is carried out, then the resulting valence state can be given as: $La^{3+}{}_{1-x}Sr^{2+}{}_xMn^{(3+x)+}O^{2-}{}_3$. Such chemical doping is an important tool in controlling the compound's physical properties. From the above example itself we can explain this. $LaMnO_3$ is a Mott insulator because of the strong on-site Coulomb interaction at Mn^{3+} site ions. However, Sr^{2+} substitution at La^{3+} site makes the $La_{1-x}Sr_xMnO_3$ perovskite conducting [7, 8]. The substitution effects change in the B-site (here Mn-site) valence, strongly influencing magnetic properties. The itinerant holes generated by the result of A-site substitution mediates double-exchange interaction between Mn^{3+} and Mn^{4+} ions which results in unusual magnetic and transport properties, including the widely studied, and technologically important, colossal magnetoresistance (CMR) effect [9, 10].

2.1 *Crystal Structure of Perovskite and Its Deviations Leading to Non-perovskite ABO₃ Oxides*

Before we embark on understanding the crystal structure and physical properties of ABO_3 type non-perovskites, let us first understand the crystal structure of a simple ABO_3 (e.g., $CaTiO_3$) perovskite.

In Fig. 1, an ideal perovskite ABX_3, structure of $CaTiO_3$, that is, the cubic structure with space group $Pm\overline{3}m$ is shown. All structures shown in this chapter have been drawn using the VESTA software—a freely available 3D visualization

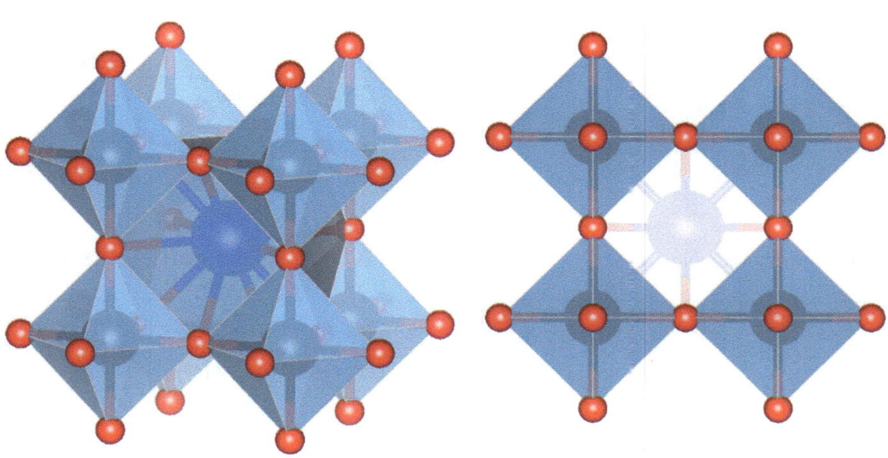

Fig. 1 (Left) Standard orientation of the $CaTiO_3$ crystal structure. (Right) View seen along c-axis is shown

program for structural models [11]. The structure can be understood as follows: The compound $CaTiO_3$ has two cations, Ca^{2+}(*A*) and Ti^{4+}(*B*) and one anion, O^{2-}(X). Each of these ions have their well-defined positions in the unit cell, where Ca^{2+} occupies the Wyckoff site 1*b*, Ti^{4+} is at 1*a* and O^{2-} is at 3*d*. The Ti^{4+}cations have six coordination by O anions, forming the TiO_6 (BO_6) octahedra. The Ca ions are seen at the corner of the unit cell, whereas the Ti ion is located at the centre of the unit cell, both the ions are connected via the oxygen ions. The typical bond distances in this structure are mentioned in Table 2.

As the TiO_6 octahedra in $CaTiO_3$ (shown via Ti-O-Ti bond angle and Ti-O bond length), represents the fact that the BO_6 octahedra plays the most important role in determining important physical properties such as magnetic and ferroelectric responses to external fields.

It is also observed that due to the synthesis conditions also, there can be variations in the BO_6 octahedra, yielding different crystal symmetry, and hence different physical properties. For example, $CaTiO_3$ synthesized at temperatures higher than 1350 °C forms in the high symmetry cubic structure, space group, $Pm\bar{3}m$. However, when the same composition is synthesized at temperatures lower than 1100 °C, the compound forms in orthorhombic structure, space group *Pcmn* (also in *Pbmn* or *Pnma*).

Table 2 shows the differences in typical bond-lengths and angles with respect to the structure in which $CaTiO_3$ forms. In Fig. 2 the qualitative phase diagram of $CaTiO_3$ is given as a representation on the close relationship between synthesis conditions and crystalline structure. Detailed information about the perovskite structures can be found in [6]. Revisiting the ideal perovskite structure, i.e., cubic structure of say, $CaTiO_3$ (space group $Pm\bar{3}m$). The structure is composed of three

Table 2 Values of bond lengths and bond angles of important bonds in $CaTiO_3$ for different structure types is compared here

Compound	Bond type	Bond length (Å)	Bond	Bond angle (°)	Source
$CaTiO_3$ (cubic)	Ti-O	1.8975(0)	Ti-O-Ti	180(0)	ICSD 31865
Space group: $Pm\bar{3}m$	Ti-Ti	3.7950(0)	O-Ca-O	90(0)	
	Ca-O	2.6834(0)			
$CaTiO_3$ (orthorhombic)	Ti-O	1.9531(6)	Ti-O-Ti	156.25(16)	ICSD 37263
Space group: *Pcmn*	Ti-Ti	3.8226(3)			
	Ca-O	2.363(5)	O-Ca-O	86.61(14)	
$CaTiO_3$ (orthorhombic)	Ti-O	1.961(3)	Ti-O1-Ti Ti-O2-Ti	156.19(19) 155.46(15)	ICSD 19790
Space group: *Pbnm*	Ti-Ti	3.8254(3)			
	Ca-O	2.4702(14)	O1-Ca-O1 O2-Ca-O2	87.06(8) 165.41(7)	

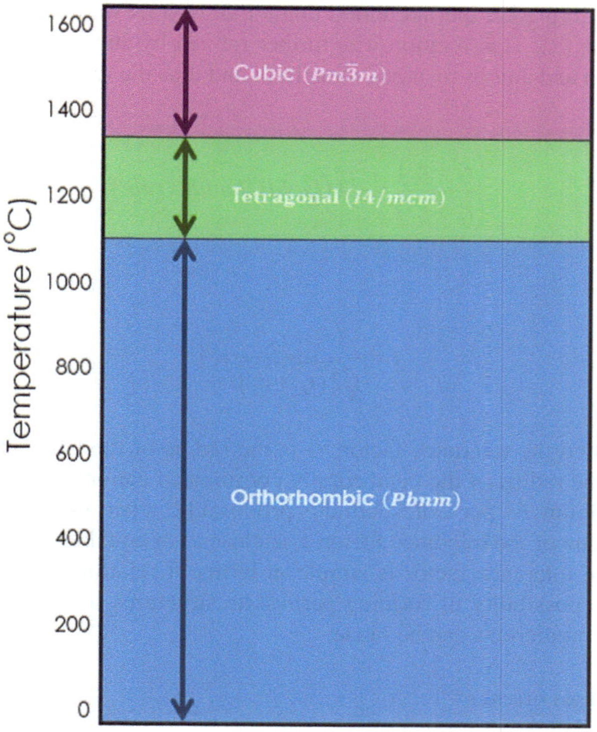

Fig. 2 Qualitative phase diagram for $CaTiO_3$

crystallographic distinct sites (Wyckoff positions) of two cations (Ca^{2+}, Ti^{4+}) and one anion (O^{2-}), where the atomic positions are given as:

Atom	x	y	z	Site	Symmetry
Ca	½	½	½	1*b*	$m-3m$
Ti	0	0	0	1*a*	$m-3m$
O	½	0	0	3*d*	$4/mm.\ m$

In this structure, as seen from the above table, the unit cell has no adjustable atomic position parameter, making it inflexible. Thus, the only way to accommodate any compositional change (substitution or change in cation/anion) can be accomplished by change in the lattice parameter. The cell parameter is given as the simple sum of anion-cation bond lengths. In the ideal perovskite unit cell, the unit cell edge, a, is equal to the twice the B-O bond length, i.e., $2(B\text{-O}) = a$; and the width of the cuboctahedral cage site, $\sqrt{2}a$, is equal to the twice the A-O bond length, i.e., $2(A\text{-O}) = \sqrt{2}$. Using these relations, Goldschmidt suggested that the ratio of the bond lengths given as:

$$\frac{(A-O)}{\sqrt{2(B-O)}}=1$$

could be used to predict the likelihood that a pair of ions would form a perovskite structure phase [2]. This relation was further refined by substituting bond-lengths between cation and anions by using ionic radii and thus the resulting relation can be written as:

$$\frac{(r_A + r_O)}{(r_B + r_O)} = \sqrt{2}\,(Goldschmidt's\,rule)$$

or

$$t = \frac{(r_A + r_O)}{\sqrt{2(r_B + r_O)}} = 1$$

where t is called the tolerance factor, r_A is the radius of the cage site cation (i.e., A-site), r_B is the radius of the octahedrally coordinated cation (B-site) and r_O is the radius of the anion. As per Goldschmidt's proposal [2] a perovskite structure would form if the value of the tolerance factor, t, is close to or equal to 1.

Though the tolerance factor is simple in terms of relationship, it does help in predicting the possibility of forming perovskite structures. Based on the t-values, the following predictions can be made:

t-values	Predicted structure
0.9–1.0	Cubic
>1.0	Hexagonal structure $r_A > r_B$; hexagonal packing of AO_3 layers
0.71–0.9	The B-site octahedra distorts close to cuboctahedral coordination polyhedron, resulting in lowering of symmetry than cubic (such as orthorhombic)
<0.71	$r_A \sim r_B$, structures such as ilmenite, $FeTiO_3$ or C-type rare earth Ln_2O_3 structures are possible

The B-site cation, which forms the BO_6 octahedra, plays an important role in controlling the physical properties of perovskites such as magnetic, ferroelectric and their responses in external magnetic fields. The electron configuration of B-cations in the BO_6 octahedra, which itself are modified by the surrounding geometry of six anions, mediate in determining these physical properties. The simplest change which can be thought of is the displacement of cations, which are usually small for the octahedral site, away from the centre of the octahedron, resulting in, among other reasons, reduction of the tolerance factor significantly less than 1 (say <0.71). With these changes, the structural symmetry reduces from cubic to tetragonal, trigonal or orthorhombic, depending upon the direction and magnitude of the displacements. The displacement of cations results in creation of permanent electric dipoles in the unit cell and can give rise to several electric properties such as pyroelectric, ferroelectric and antiferroelectric effects [6]. When A-site cation's ionic radii are too small for cuboctahedral cage site, BO_6 octahedra twist to effectively reduce the cavity dimensions, which allows to accommodate cations even if $t < 1$. As seen above, the cation displacement results in reduction of symmetry, the rotation of BO_6 octahedra also results in lowering of the symmetry. Octahedral tilt or rotation is another structural response which helps in preserving the perfect BO_6 octahedral geometry. The third way in which the perovskite structure can change to perovskite-like structure is when distortion of BO_6 octahedron takes place, in rare cases, to give elongated or flattened octahedra, which can also lead to square planar or square pyramidal coordination. Such effects can take place due to the interactions between the electrons of cations and surrounding anions, usually understood as Jahn-Teller effect [12]. Due to the disproportionation of cation valence changes octahedral distortion can take place: $2B^n \rightarrow B^{n+1} + B^{n-1}$. The two different-sized cations may adjust to the surroundings by distorting one or both BO_6 octahedra, giving rise to two different-sized octahedra. These three modifications, i.e., displacement of B-cation, tilt/rotation of BO_6 octahedra and distortion of BO_6, can occur independently or in combination with one another. Though these effects may cancel each other, they do have influence on microscopic properties. These distortions are usually small and can be influenced by ambient conditions such as changes in temperature, crystallite size or form, pressure may alter the degree of distortion or the type of distortion. As seen in the case of $CaTiO_3$ (Fig. 2), most perovskite phases manifest a series of symmetry changes as the temperature or pressure is changed [6].

Now moving on to non-perovskite ABO_3 compounds, the difference between structural properties must be understood first. There are several such compounds which have ABO_3 stoichiometry but are structurally different from perovskites. Some well-studied compounds among this class are the ilmenite (e.g., $FeTiO_3$), bixbyite (e.g., Mn_2O_3, $FeMnO_3$) etc. We shall be discussing the physical properties of these compounds and compare them with compounds forming in perovskite structures.

2.2 *Crystal Structures of Non-perovskite Compounds, $FeTiO_3$ (Ilmenite) and $FeMnO_3$ (Bixbyite) and Its Comparison with Perovskite ABO_3 Oxides*

In general, ABO_3 perovskite oxides have A-site which have larger cations and are 12-fold coordinated, whereas B-site has smaller cation (compared to A-site) which have sixfold coordinated BO_6 octahedra, which are corner sharing. Some famous ABO_3 type perovskites are $CaTiO_3$, $SrTiO_3$, $LaMnO_3$, $BaTiO_3$, $YFeO_3$ etc. Some key properties of the perovskite compounds are high tunability via doping or pressure, exhibits interesting physical properties such as ferroelectricity, multiferroicity, superconductivity, colossal magnetoresistance, ionic conductivity etc. The usability of perovskite compounds is found in fuel cells, sensors, photovoltaics etc.

The main difference between perovskite and non-perovskite ABO_3 compounds is in the difference in coordination and connectivity of cations/anions. In non-perovskites such as Ilmenite (trigonal system, *R*-3 space group), the *B*-site cations are edge-sharing octahedra, which forms a layered honeycomb network. This edge-sharing connectivity influences the electronic and magnetic properties making them useful for spintronic applications, whereas in bixbyite structure, BO_6 octahedra are primarily corner-sharing, forming a cubic structure in $Ia\overline{3}$ space group. This arrangement leads to slightly distorted fluorite-like lattice, where one-fourth of the oxygen sites are vacant, influencing its electronic and magnetic properties.

The structural features of $FeTiO_3$ and $FeMnO_3$ are shown in Figs. 3 and 4 respectively. The ilmenite $FeTiO_3$ is displayed in Fig. 3 in different orientations. The compound crystallized in trigonal structure (space group *R*-3) which is an ordered derivative of Al_2O_3 type corundum framework. In this structure, Fe^{2+} and Ti^{4+} ions are placed in alternate layers, perpendicular to the trigonal *c*-axis [13]. Here, both Fe and Ti ions are surrounded by six oxygen atoms, forming the BO_6 octahedra. The

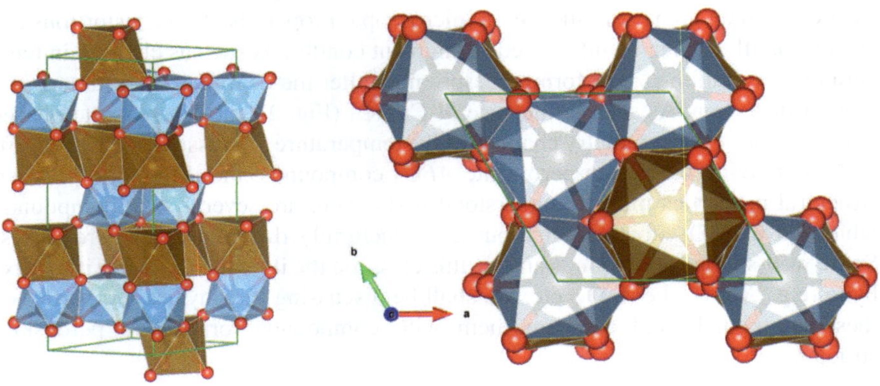

Fig. 3 (Left) Standard orientation of the $FeTiO_3$ crystal structure. (Right) View seen along c-axis is shown

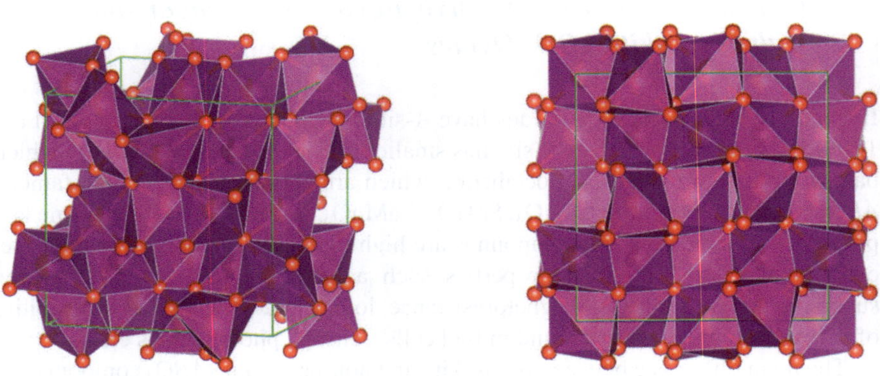

Fig. 4 (Left) Standard orientation of the $FeMnO_3$ crystal structure. (Right) View seen along c-axis is shown

main distinction between ilmenite and perovskite structure is the edge-sharing octahedra in the former, which creates a honeycomb-line arrangement (Fig. 3). Owing to the trigonal symmetry the close-packing of oxygen atoms, contributes to anisotropic electric and magnetic properties of $FeTiO_3$. Some known examples of ilmenite structures are $CdTiO_3$, $CoTiO_3$, $MgTiO_3$, $MgGeO_3$, $LiNbO_3$, $CrRhO_3$, $FeRhO_3$, $FeVO_3$, etc.

The compound iron manganite ($FeMnO_3$) crystallizes in bixbyite structure, space group $Ia\overline{3}$. Two different orientations of the bixbyite structure of $FeMnO_3$ are shown in Fig. 4. This structure is a slightly distorted version of the fluorite lattice, where Fe and Mn occupies both 8*b* and 24*d* Wycoff sites in 1:1 ratio [14]. Here, both Fe and Mn ions are surrounded by six oxygen atoms, forming BO_6 octahedra, which are corner-sharing octahedra contributing to the bixbyite framework. The presence of magnetic Fe and Mn at both sites strongly influences the magnetism and electronic properties of $FeMnO_3$ and this compound finds applications in frustrated magnetism, catalysis, energy storage etc. [14, 15]. A table highlighting the main differences/properties of perovskite and non-perovskite systems is given Table 3.

2.3 *Physical Properties of Non-perovskite Compounds, $FeTiO_3$ (Ilmenite) and $FeMnO_3$ (Bixbyite)*

2.3.1 Magnetic Structure and Properties

Comparing magnetic properties of perovskite and non-perovskite oxides, say, (Ba/Sr/La)FeO_3 and Fe(Mn/Ti)O_3 compounds [14, 16–18], one can find that all the three-perovskites does not show long-range magnetic ordering below room temperature. However, some of these structures show ferromagnetism, antiferromagnetism, ferrimagnetism etc., depending upon the choice of ions at *A* and *B* site and the structure they form into. Starting with the magnetic properties of $LaFeO_3$, also

Table 3 Key structural differences between perovskite and non-perovskite structures

Feature	Perovskite	Non-perovskite
BO_6 (Octahedra) connectivity	Corner-sharing	Edge/face-sharing or distorted network
A-site coordination	12-fold	Varies (6–12)
Stability condition	Tolerance factor (~0.8 to 1.0)	Outside ideal tolerance factor range
Phase complexity	Can support phase coexistence	Often chemically or topologically distinct
Why study?	Structurally flexible, supports complex interactions between spin, charge, orbital and lattice degrees of freedom	Offers distinct platforms for magnetic frustration, low-dimensional behaviour, ionic selectivity etc.

known as lanthanum orthoferrites, the magnetic properties of bulk (polycrystalline) sample show antiferromagnetic ordering (T_N) at an elevated temperature of 740 K, however, in nanoscale sized sample the increased surface area leads to uncompensated surface spins resulting in weak ferromagnetic behaviour. It is also reported that defects like oxygen vacancies or distortions in the FeO_6 octahedra can induce spin canting, which also contributes to ferromagnetism [19, 20]. $SrFeO_3$ exhibits paramagnetic to magnetic ordering at around 134 K, where it transits into a helical magnetic order [21, 22]. $BaFeO_3$ is an interesting perovskite sample, as it can exist in different structural phases depending upon how it is prepared. $BaFeO_3$ in hexagonal structural phase undergoes transition from paramagnetic to ferromagnetic ordering around 250 K and undergoes another magnetic transition to antiferromagnetic state below 160 K. In the cubic form $BaFeO_3$, undergoes a paramagnet to spin-spiral antiferromagnetic transition around 111 K. Oxygen deficiency is usually observed in perovskite materials and that can also alter the magnetic properties, such as in case of $BaFeO_3$, oxygen deficiency can induce ferromagnetism [23–25].

Comparing the magnetic properties of iron-based perovskites with that of iron based non-perovskites, we find that $FeMnO_3$ exhibits complex magnetic behaviour where it exhibits ferrimagnetism at room temperature and antiferromagnetic order (T_N) below 36 K. Some features of magnetic structure of $FeMnO_3$ can be listed as follows:

- Paramagnetic space group: $Ia\overline{3}$
- Magnetic structure from MAGNDATA database (#0.508)
- Propagation vector: k1 (0, 0, 0)
- BNS Magnetic space group: $Ib'c'a$ (#73.551)
- Magnetic Point Group: $m'm'm$ (8.4.27)

This compound also exhibits significant magnetocaloric property (change in magnetic entropy) around its Néel temperature [26]. The chemical structure of $FeMnO_3$ is unique in the sense that Fe and Mn occupy both *A* and *B* site in equal proportions. The magnetic structure of $FeMnO_3$ is shown in Fig. 5 [27]. Magnetic structure of $FeMnO_3$ studied using neutron diffraction measurements show that the Fe and Mn moments at *A* and *B* sites order in anti-parallel arrangements, giving rise to overall ferrimagnetic order at room temperature. When the sample is cooled below room temperature, it undergoes antiferromagnetic ordering at around 36 K, converting the moments at *A* and *B* site equal, but oppositely aligned.

The ilmenite $FeTiO_3$, exhibits paramagnet to antiferromagnetic transition (T_N) ~55 to 60 K, and further at low temperatures it also exhibits weak ferromagnetism owing to surface effects or cation ordering [31]. The ilmenite structure of $FeTiO_3$ is derived from the rhombohedral structure is derived from α-Fe_2O_3, where one Fe^{3+} ion is replaced by Ti^{4+}.

In the magnetic structure (Fig. 6), every alternate Fe atom slightly buckles along the c-axis and the spin directed along the c-axis exhibits tilting of about 2°. Below T_N, $FeTiO_3$ exhibits antiferromagnetic ordering, meaning that neighbouring magnetic moments of Fe ions align in opposite directions arranged in alternating layers along the c-axis, with spins pointing up in one layer and down in the adjacent layer

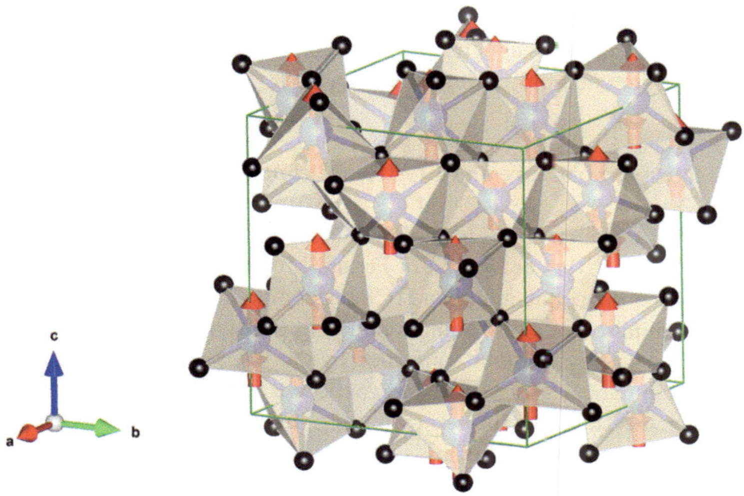

Fig. 5 Room temperature magnetic structure of $FeMnO_3$ determined from neutron diffraction data [27]. Structure drawn in VESTA [11] using mCIF (#0.508_FeMnO3) from MAGNDATA [28–30]

Fig. 6 Magnetic structure of $FeTiO_3$ [32]. Only Fe spins are shown, which are pointed along c-axis. Structure drawn using CIF file of ICSD # 187688 using VESTA software [11]

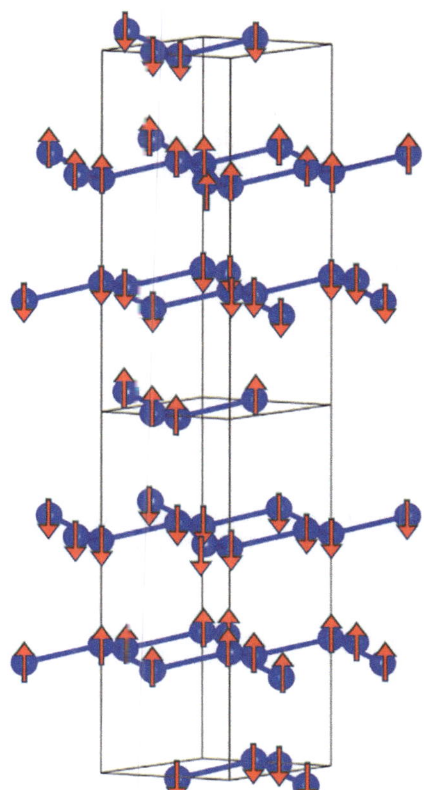

[32, 33]. $FeTiO_3$ is also known to exhibit metamagnetic transition, i.e., field induced cross over from antiferromagnetic to ferromagnetic state, resulting in all Fe spins aligned in one direction only along the c-axis [34].

2.3.2 Electronic Structure and Electrical Properties

The band structure calculations for $FeMnO_3$ using DFT calculations [14] shows that the valence band (VB) is composed of Fe and Mn (3*d*) orbitals hybridized with O (2*p*) states while the conduction band (CB) is composed mainly of empty states of Fe and Mn orbitals, leading to semi-conducting behaviour of $FeMnO_3$. The semi-conducting behaviour is further confirmed by the values of spin-polarization. The magnetic structure of $FeMnO_3$ is ferrimagnetic and hence the VB and CB are asymmetrical, which results in spin-gap exhibiting higher excitation energy ~3.81 eV, which is akin to the semiconducting system [14]. Therefore, in this compound the electrical transport depends on the distribution of cations, enabling hopping of the charge carriers (electrons or holes due to $Mn^{3+/4+}$ and $Fe^{2+/3+}$), and presence of oxygen vacancies. The combination of the oxygen vacancies and hopping of charge carriers influences the electric transport in this compound. $FeMnO_3$ is also used/studied as potential anode material for lithium-ion batteries due to its higher theoretical capacity for lithium storage [35].

The compound $FeTiO_3$ is known as the wide band gap semiconductor with a band gap of ~2 eV. Its electronic structure is characterized by a wide band gap between valence band, mainly composed of Fe (3*d*) and O (2*p*) orbitals, and conduction band, made up of Ti (3*d*) orbitals. The strong hybridization between Ti (3d) and O (2*p*), plays pivotal role in charge transfer. Also, the presence of Fe^{2+} and Ti^{4+} leads to local and non-local intervalence state, resulting in charge transfer [36–38]. $FeTiO_3$ is found to be exhibiting *p*-type conductivity, which is attributed to oxygen vacancies [39].

2.3.3 Dielectric Properties

The dielectric behaviour of non-perovskite oxides such as $FeMnO_3$ (bixbyite) and $FeTiO_3$ (ilmenite) is significantly influenced by their unique crystal structure, defect chemistry, and charge transport mechanisms. Unlike classic perovskite oxides like $BaTiO_3$, which show strong ferroelectric transitions and extremely high dielectric constants ($\varepsilon > 1000$ near the Curie point), these non-perovskites typically exhibit modest dielectric constants and relaxation-type behaviour, reflecting their semiconducting and defect-mediated nature.

$FeMnO_3$ (Bixbyite)

$FeMnO_3$ exhibits frequency dependent dielectric behaviour characterized by a gradual dispersion in permittivity (ε′) at low frequencies (<100 Hz), primarily due to interfacial polarization mechanisms, such as Maxwell–Wagner effects. These effects are typically associated with charge accumulation at the grain boundaries and defect interfaces. At higher frequencies, dielectric response is increasingly dominated by atomic polarization and localized electronic dipole reorientation [15].

The presence of oxygen vacancies and mixed valence states (Fe^{3+}/Mn^{4+} or Fe^{2+}/Mn^{3+}) promotes hopping conduction, leading to dielectric loss (tan δ) especially at elevated temperatures. These features indicate that the material exhibits relaxor-like behaviour rather than sharp ferroelectric transitions.

Such behaviour positions $FeMnO_3$ as a candidate for energy storage capacitors or dielectric layers in thermally robust electronics, though its performance remains lower than conventional ferroelectrics.

$FeTiO_3$ (Ilmenite)

$FeTiO_3$ exhibits a more complex dielectric profile. Studies have shown the presence of two dielectric relaxations occurring at approximately 52 and 118 K, attributed to electron hopping ($Fe^{2+} \rightleftharpoons Fe^{3+}$) and interfacial polarization due to grain boundaries and lattice distortions. These relaxations are thermally activated and display Arrhenius-type behaviour, consistent with defect-mediated mechanisms.

Interestingly, $FeTiO_3$ shows high dielectric losses at temperatures above 200 K due to an increase in intrinsic conductivity, making the accurate determination of ε at high temperature challenging. This conductivity is typically *p*-type, driven by oxygen vacancies and charge delocalization across Fe–Ti–O bonds. The dielectric behaviour is also found to be sensitive to synthesis methods, crystallinity, and defect concentration [37, 40, 41].

Despite the absence of spontaneous polarization due to the centrosymmetric trigonal structure, some off-centering of cations or grain boundary effects may induce weak local polarizations, which can mimic ferroelectric-like anomalies in nanoscale systems.

2.3.4 Thermal Transport Properties

Thermal transport in non-perovskite oxides is primarily governed by phonon scattering mechanisms, though magnetic and electronic contributions become relevant in magnetically ordered or semiconducting systems.

$FeMnO_3$ (Bixbyite)

$FeMnO_3$, with its magnetic ordering and vacancy-induced lattice distortions, demonstrates strong phonon–magnon interactions, especially near its Néel temperature (~36 K). These interactions modulate the heat capacity and thermal conductivity due to the scattering of phonons by spin fluctuations.

Additionally, its magnetocaloric effect (MCE)—the reversible temperature change under an applied magnetic field—has been reported to peak around T_N, highlighting coupling between magnetic entropy and thermal response. This makes $FeMnO_3$ a candidate for magnetic refrigeration applications, especially at cryogenic temperatures [26]. The magnetocaloric properties exhibited by $FeMnO_3$ as a function of temperature also is an indication of the interesting thermal transport properties [42–44].

From a structural perspective, the bixbyite lattice, with its ordered oxygen vacancies, introduces scattering centres for phonons, thereby lowering the lattice thermal conductivity (κ). This suppressed κ can be beneficial for thermoelectric applications, provided the electrical conductivity can be appropriately tuned.

$FeTiO_3$ (Ilmenite)

$FeTiO_3$ exhibits relatively low thermal conductivity (~2.8 W/mK at room temperature), comparable to other oxide ceramics. Its thermal transport is largely phonon-limited, and deviations from ideal behaviour arise due to cation disorder, octahedral tilts, and electron–phonon coupling [36, 39].

Specific heat measurements reveal anomalies near the antiferromagnetic transition ($T_N \approx 55$–60 K), indicating coupling between magnetic and lattice degrees of freedom. At higher temperatures (~642 K), DSC (Differential Scanning Calorimetry) studies show exothermic peaks that correspond to structural transitions or oxidation of Fe^{2+} to Fe^{3+}, associated with abrupt changes in unit cell volume and entropy. This sensitivity to redox chemistry implies that $FeTiO_3$ can also act as a functional material in high-temperature thermochemical cycles, potentially useful for oxygen storage or catalytic applications [45].

3 Discussions and Conclusions

Based on the discussions in the above passages, it is clearly evident that physical properties of materials and their applicability for devices strongly depends upon the constituent elements (magnetic or non-magnetic ions), structural properties and external stimuli's such as temperature, pressure, electrical or magnetic fields. In Table 4, comparison of structural properties is listed. Drawing comparison between well-studied perovskite compounds and not so well studied, ilmenite ($FeTiO_3$) and bixbyite ($FeMnO_3$) type compounds is relevant here to highlight the potential of these non-perovskite compounds in understanding basic physics and also their application potential.

Table 4 Comparison of structural properties of perovskite and non-perovskites

Feature	Perovskite (ABO_3)	Ilmenite ($FeTiO_3$)	Bixbyite ($FeMnO_3$)
Crystal system	Cubic (often distorted)	Trigonal (rhombohedral)	Cubic
Space group	$Pm\overline{3}m$	$R\overline{3}$	$Ia\overline{3}$
General stoichiometry	ABO_3	$AB'O_3$	Mn_2O_3 (but often written as $FeMnO_3$)
Cation coordination	A:12-fold; B: sixfold	A: sixfold (Fe^{2+}); B: sixfold (Ti^{4+})	Fe^{3+}/Mn^{3+}: sixfold (octahedral)
Anion coordination	O in octahedral framework	O links Fe and Ti octahedra	O: coordinated to multiple cations
Key features	Corner-sharing BO_6 octahedra	Alternating FeO_6 and TiO_6 layers	Ordered oxygen vacancies
Vacancies	None in ideal form	None	1/6 of O sites vacant (~2 per unit)
Common properties	Ferroelectric, multiferroic	Weakly magnetic, insulating	Magnetic, catalytic, semiconducting

The physical properties of perovskites consist of 3D network of corner-sharing BO_6 octahedra with A-site cations in large interstices. Distortions of the usual cubic structure leads to orthorhombic or tetragonal symmetries and in rare-cases to rhombohedral/hexagonal structures. Due to these structural features, perovskite oxides of ABO_3 type are most versatile structures and widely studied. Ilmenite structure such as $FeTiO_3$ are layered structures with Fe^{2+} and Ti^{4+} in octahedral coordination placed in alternating cation planes. The A-site lacks 12-fold coordination, thus deviating from the perovskite structures. Bixbyite structure on the other hand are derived from the fluorite (CaF_2) structure, but with ordered oxygen vacancies. The removal of oxygen i.e., creation of anion vacancies is not random, but these vacancies are ordered in a specific way (vacancy ordering), thus leaving the cations surrounded by distorted octahedra (due to anion vacancy). In Table 5, comparison of physical properties of few selected examples of perovskite and non-perovskite compounds is given.

The main feature of these three classes of compounds is that perovskites are the most versatile compounds as they exhibit variety of physical properties such as ferroelectrics, has tuneable band-gaps, has conductivity over wide temperature range and high dielectric constant, exhibits variety of magnetic properties, which makes them suitable for applications in sensors, actuators, solar cells, memories (electronic devices etc.). However, rare-earth free magnetic materials which are easy to fabricate, having tremendous application potential are forming the basis for materials for new generation of electronics and devices. In view of this, the ilmenite compounds, having wide-band gap semiconductors, antiferromagnet, are useful in variety of applications such as pigments, photocatalysis, etc. The bixbyite compounds having oxygen-vacancy enable conduction have tremendous applications in catalysis, and as anode material in lithium-ion batteries. Being magnetic at room temperature also makes them attractive option for ferrimagnet based spintronics and sensor applications.

Table 5 Comparative perspective and application relevance

Feature	Perovskite (*ABO*$_3$ e.g., BaTiO$_3$)	Ilmenite (FeTiO$_3$)	Bixbyite (FeMnO$_3$)
Density (g/cm^3)	~4 (CaTiO$_3$); ~6 (BaTiO$_3$)	~4.7 to 4.8	~4.5 to 5.0
Band gap	~3.2 eV	~2.6 eV (wide-gap semiconductor)	~2.0 to 3.0 eV (semiconducting, mixed Fe^{3+}/Mn^{4+} states)
Magnetism	Generally non-magnetic (BaTiO$_3$); exceptions when *B*-site = transition metal ion	Antiferromagnetic, weak magnetism at low T	Antiferromagnetic or ferrimagnetic at low temperatures
Ferroelectricity	Strong in oxides like BaTiO$_3$; spontaneous polarization via B-site displacement	None (centrosymmetric)	Generally, none (cubic), though vacancy distortions may create local dipoles
Electrical conductivity	Wide range (insulating → conductive oxides like LSC, LSM: 10^2–10^3 S/cm)	Low (semiconducting)	Moderate (oxygen vacancy mediated conduction)
Thermal conductivity	Variable (~2 to 6 W/mK)	~2.8 W/mK	Low-moderate (~1 to 2 W/mK)
Specific heat anomaly	At ferroelectric transition temperature (~393 K for BaTiO$_3$)	Near T_N and 642 K	Near T_N (~36 K)
Dielectric constant (ε′)	Very high (BaTiO$_3$ ≫ 1000 near T_C)	Moderate (<100)	Moderate (50–200)
Relaxation	Sharp transition (ferroelectric)	Two relaxations (e^- hopping)	Maxwell-Wagner hopping
Mechanical hardness	Brittle ceramics; Mohs around 5–6, anisotropic elastic moduli	Similar to dense ceramics (~6 to 7 Mohs)	Comparable to other sesquioxides (~5 to 6 Mohs)
Application potential	Sensors, actuators, thermoelectrics	Pigments, dielectrics, catalysis	Spintronics, Li-ion anodes

References

1. De Graef, M., McHenry, M.E.: Structure of Materials—An Introduction to Crystallography, Diffraction and Symmetry, 2nd edn. Cambridge University Press, Cambridge (2012)
2. Kay, H.F., Bailey, P.C.: Structure and properties of CaTiO$_3$. Acta Crystallogr. **10**, 219–226 (1957)
3. Goldschmidt, V.M.: Die Gesetze der Krystallochemie. Naturwissenschaften. **14**, 477–485 (1926). https://doi.org/10.1007/BF01507527
4. Sajjad, M., Mahmood, Q., Singh, N., Andreas Larsson, J.: Ultralow lattice thermal conductivity in double perovskite Cs$_2$PtI$_6$: promising thermoelectric material. ACS Appl. Energy. **3**
5. Shimakawa, Y.: Chapter 6—Perovskite structure compounds. In: Dronskowski, R., Kikkawa, S., Stein, A. (eds.) Handbook of Solid-State Chemistry, 1st edn, pp. 221–250. Wiley-VCH GmbH & Co. KGaA, Weinheim (2017)
6. Tilley, R.J.D.: Perovskites: Structure-Property Relationships. John Wiley & Sons, Ltd., Chichester (2016)
7. Zener, C.: Interaction between the d-shells in the transition metals. II. Ferromagnetic compounds of manganese with perovskite structure. Phys. Rev. B. **82**, 403–405 (1951)
8. de Gennes, P.G.: Effects of double exchange in magnetic crystals. Phys. Rev. **118**, 141 (1960)

9. Tokura, Y., Urushibara, A., Moritomo, Y., Arima, T., Asamitsu, A., Kido, G., Furukawa, N.: Giant magnetotransport phenomena in filling-controlled Kondo lattice system: La1-xSrxMnO3. J. Phys. Soc. Jpn. **63**, 3931–3935 (1994)
10. Ramirez, A.P.: Colossal magnetoresistance. J. Phys. Condens. Matter. **9**, 8171–8199 (1997)
11. Momma, K., Izumi, F.: VESTA 3 for three-dimensional visualization of crystal, volumetric and morphology data. J. Appl. Crystallogr. **44**, 1272–1276 (2011)
12. O'Brien, M.C.M., Chancey, C.C.: The Jahn-Teller effect: an introduction and current review. Am. J. Phys. **61**, 688–697 (1993)
13. Mehl, M.J., Hicks, D., Toher, C., Levy, O., Hanson, R.M., Hart, G., Curtarolo, S.: The AFLOW library of crystallographic prototypes: part 1. Comput. Mater. Sci. **136**, S1–S828 (2017)
14. Rayaprol, S., Ribeiro, R.A.P., Singh, K., Reddy, V.R., Kaushik, S.D., de Lazaro, S.R.: Experimental and theoretical interpretation of magnetic ground state of $FeMnO_3$. J. Alloys Compd. **774**, 290–298 (2019)
15. Lobo, L.S., Rubankumar, A.: Investigation on structural and electrical properties of $FeMnO_3$ synthesized by sol-gel method. Ionics. **25**, 1341–1350 (2019)
16. Hong, N.H., Kanoun, M.B., Kim, J.-G., Sh. Atabaev, T., Konishi, K., Noguchi, S., Kurisu, M., Goumri-Said, S.: Shaping the magnetic properties of $BaFeO_3$ perovskite-type by alkaline-earth doping. J. Phys. Chem. C. **122**, 2983–2989 (2018)
17. Acharya, S., Mondal, J., Ghosh, S., Roy, S.K., Chakrabarti, P.K.: Multiferroic behaviour of lanthanum orthoferrite ($LaFeO_3$). Mater. Lett. **64**, 415–418 (2010)
18. Ohara, S., Naka, T., Sunakawa, K., Kubuki, S., Senna, M., Hashishin, T.: Emergence of ferromagnetism due to charge transfer in compressed ilmenite powder using super-high-energy ball milling. Sci. Rep. **10**, 5293 (2020)
19. Lee, W.-Y., Yun, H.J., Yoon, J.-W.: Characterization and magnetic properties of $LaFeO_3$ nanofibers synthesized by electrospinning. J. Alloys Compd. **583**, 320–324 (2014)
20. Chudinovych, O.V., Vedel, D.V., Stasyuk, O.O., Tomila, T.V., Aguirre, M.H.: Production, structure and magnetic properties of nanocomposites based on the perovskite phase $LaFeO_3$. Solid State Sci. **157**, 107699 (2024)
21. Kawanaka, H., Kawawa, E., Nishihara, Y., Iga, F., Kondo, A., Kindo, K., Matsuda, Y.: Magnetic properties of perovskite $Ca_{1-x}Sr_xFeO_3$. AIP Adv. **8**, 101418 (2018)
22. Takeda, T., Yamaguchi, Y., Watanabe, H.: Magnetic structure of $SrFeO_3$. J. Phys. Soc. Jpn. **33**, 967–969 (1972)
23. Iga, F., Nishihara, Y., Katayama, T., Murata, K., Takeda, Y.: Magnetic and transport properties of $BaFeO_{3-y}$. J. Magn. Magn. Mater. **104–107**, 1973–1975 (1992)
24. Kawazoe, Y., Kanomata, T., Note, R.: $BaFeO_3$ in High Pressure Materials Properties: Magnetic Properties of Oxides Under Pressure. Springer, Berlin (2023). https://doi.org/10.1007/978-3-662-64593-2_119
25. Gutiérrez-Martín, D., Hernando, M., Parras, M., Torres-Pardo, A., Fernández-Díaz, M.T., Savvin, S., González-Calbet, J.M., Rodrigues, J.E., Silva Jr., R.S., Martínez, J.L., Varela, A.: Unveiling the magnetic structure of $BaFeO_{3-y}$: shedding light on the elusive magnetic behaviour. J. Alloys Compd. **1010**, 177081 (2025)
26. Rayaprol, S., Kaushik, S.D.: Magnetic and magnetocaloric properties of $FeMnO_3$. Ceram. Int. **41**, 9567–9571 (2015)
27. Rayaprol, S., Kaushik, S.D., Babu, P.D., Siruguri, V.: Structure and magnetism of $FeMnO_3$. AIP Conf. Proc. **1512**, 1132–1133 (2013)
28. Gallego, S.V., Manuel Perez-Mato, J., Elcoro, L., Tasci, E.S., Hanson, R.M., Momma, K., Aroyo, M.I., Madariaga, G.: MAGNDATA: towards a database of magnetic structures—I, the commensurate case. J. Appl. Crystallogr. **49**, 1750–1776 (2016)
29. Gallego, S.V., Manuel Perez-Mato, J., Elcoro, L., Tasci, E.S., Hanson, R.M., Aroyo, M.I., Madariaga, G.: MAGNDATA: towards a database of magnetic structures. II. The incommensurate case. J. Appl. Crystallogr. **49**, 1941–1956 (2016)
30. Perez-Mato, J.M., Gallego, S.V., Tasci, E.S., Elcoro, L., de la Flor, G., Aroyo, M.I.: Symmetry-based computational tools for magnetic crystallography. Annu. Rev. Mater. Res. **45**, 13.1-13.32 (2015)

31. Srinivas, P., Sendil Kumar, A., Babu, P.D., Bhatnagar, A.K.: Synthesis and magnetic properties of nanocrystalline $FeTiO_3$ materials. J. Supercond. Novel Magn. **31**, 2017 (2018)
32. Lida, K., Kodama, K., Inamura, Y., Nakamura, M., Chang, L.-J., Shamoto, S.-i.: Magnon mode transition in real space. Sci. Rep. **12**, 20663 (2022)
33. Kato, H., Yamaguchi, Y., Ohashi, M., Yamada, M., Takei, H., Funahashi, S.: On the obliqueness of the easy axis in collinear antiferromagnetic $FeTiO_3$. Solid State Commun. **45**, 669–672 (1983)
34. Zafar, A., Milinda Abeykoon, A.M., Petkov, V.: Structural mechanism of the magnetic phase transitions in $FeTiO_3$. Phys. Rev. Mater. **8**, 093602 (2024)
35. Cao, K., Liu, H., Xu, X., Wang, Y., Jiao, L.: $FeMnO_3$: a high-performance Li-ion battery anode material. Chem. Commun. **52**, 11414–11417 (2016)
36. Chen, S.W., Huang, M.J., Lin, P.A., Jeng, H.T., Lee, J.M., Haw, S.C., Chen, S.A., Lin, H.J., Lu, K.T., Chen, D.P., Dou, S.X., Wang, X.L., Chen, J.M.: Orbital structure of $FeTiO_3$ ilmenite investigated with polarization-dependent X-ray absorption spectroscopy and band structure calculations. Appl. Phys. Lett. **102**, 042107 (2013)
37. Sheikh, A., Soni, K., Lakshmi, N.: Study of structural and electrochemical properties of ilmenite ($FeTiO_3$) doped TiO_2. Mater. Today Proc. **49**, 1857–1860 (2022)
38. Phoohinkong, W., Boonyarattanakalin, K., Mekprasart, W., Pavasupree, S., Pecharapa, W.: Nonlocal XANES pre-edge feature of $FeTiO_3$ ilmenite-type at Ti and Fe K-edge. Radiat. Phys. Chem. **174**, 108919 (2020)
39. Andreozzi, G.B., Cellucci, F., Gozzi, D.: High-temperature electrical conductivity of $FeTiO_3$ and ilmenite. J. Mater. Chem. **6**, 987–991 (1996)
40. Iwauchi, K., Kiyama, M., Nakamura, T.: Dielectric properties of $FeTiO_3$. Phys. Status Solidi (A). **127**, 567–572 (1991)
41. Parida, B.K., Bhowmik, R.N.: Structural phase, surface chemical state, optical and dielectric properties of $FeTiO_3$ at different conditions. AIP Conf. Proc. **2995**, 020109 (2024)
42. Gowreesan, S., Ruban Kumar, A.: Structural, magnetic, and electrical property of nanocrystalline perovskite structure of iron manganite ($FeMnO_3$). Appl. Phys. A Mater. Sci. Process. **123**, 689 (2017)
43. Li, Z.J., Wang, S.F., Li, B., Xiang, X.: A new method for synthesis of $FeMnO_3$ ceramics and its phase transformation. J. Nano Res. **37**, 122–131 (2015)
44. Li, S., Guo, S., Hoke, T., Chen, X.: Thermal transport in magnetic materials: a review. Mater. Today Electron. **12**, 100156 (2025)
45. Song, J., Fan, D., Huang, S., Zhang, S., Wu, M., Chen, W., Zhou, W.: Thermal expansivity of geikielite and ilmenite utilizing in-situ synchrotron X-ray diffraction at high temperature. Acta Geochim. **42**, 988–997 (2023)

Industrial Process Simulators Applied to the Study of Biomass Conversion

Lucas Bonfim-Rocha, Mariana Martini Rodrigues, Fábio Eiji Nishyiama, Sara Regina Osipi, Caliane Bastos Borba Costa, and Felipe de Almeida La Porta

Abstract This chapter examines the application of industrial process simulators in the study of biomass conversion, emphasizing their importance for developing, optimizing, and controlling industrial processes. Chemical process simulators, both in static and dynamic states, enable the analysis of complex systems without requiring physical experimentation, thus saving time and resources. Key biomass conversion technologies currently include thermochemical processes (such as gasification, pyrolysis, and combustion), biochemical processes (such as fermentation and anaerobic digestion), and physicochemical processes (such as hydrogenation and transesterification for renewable diesel production). These technologies benefit significantly from the use of simulators to optimize performance, assess economic feasibility, integrate energy systems, and analyze environmental impacts. Static simulation evaluates steady-state conditions, while dynamic simulation is essential for analyzing temporal variations in transient processes and control systems. These simulations solve systems of algebraic differential equations that describe mass and energy balances, incorporating thermodynamic and kinetic models. Predicting operational scenarios and identifying process improvements is critical for advancing sustainable and efficient biomass conversion technologies. Accurate simulations

L. Bonfim-Rocha · M. M. Rodrigues
Department of Chemical Engineering, Technological Federal University of Paraná (UTFPR), Londrina, PR, Brazil

S. R. Osipi · F. E. Nishyiama · C. B. B. Costa
Chemical Engineering Graduate Program, State University of Maringá (UEM), Maringá, PR, Brazil

F. de Almeida La Porta (✉)
Department of Chemistry, Technological Federal University of Paraná (UTFPR), Londrina, PR, Brazil

Post-Graduation Program in Chemistry, State University of Londrina, Londrina, PR, Brazil

Faculty of Informatics and Management, University of Hradec Kralove, Hradec Kralove, Czech Republic
e-mail: felipelaporta@utfpr.edu.br

C. A. Taft, S. R. d. Lazaro (eds.), *Innovative Materials for Energy and Sustainability*, Engineering Materials,
https://doi.org/10.1007/978-3-032-09149-9_4

rely on reliable data and the proper adjustment of model variables, underscoring the importance of precise instrumentation and comprehensive process information. This chapter spotlights the pivotal role of simulation tools in the ongoing development of the biomass conversion industry, addressing the growing demands for sustainability and energy efficiency.

Keywords Simulation · Industry · Biochemical processes · Bioenergy · Biofuels

1 Introduction

The transition toward sustainable industrial practices has prompted growing interest in advanced technologies that can optimize the transformation of biomass into renewable energy and commercially valuable products. In parallel, industrial process simulators have emerged as powerful tools for modeling, analyzing, and assessing complex biomass conversion processes. These computational platforms enable researchers and engineers to design, simulate, and optimize various thermochemical and biochemical pathways, ultimately facilitating the scaling-up of laboratory findings to industrial-scale operations with greater efficiency and reduced uncertainty [1].

Specifically, for biochemical processes, biomass—a renewable and widely available resource—presents a promising alternative to fossil feedstocks for producing biofuels, biochemicals, and, in addition, other bioproducts [2]. However, biomass conversion technologies are inherently multifaceted, involving numerous reactions, intermediates, and process configurations that are highly sensitive to variations in feedstock and operating conditions. Given this challenge, process simulation tools may serve as indispensable resources for predicting system behavior, assessing techno-economic and environmental performance, and guiding decision-making throughout the development pipeline [3].

In recent years, the integration of simulation tools with experimental data and life cycle assessment has enabled more accurate representations of biomass conversion systems [4]. A range of specialized software tools, known as process simulators, are employed; these include Aspen Plus, Aspen Hysys, DWSIM, SuperPro Designer, among others, to support the modeling of processes including gasification, pyrolysis, hydrothermal liquefaction, fermentation, and catalytic processes. Such tools enable the evaluation of material and energy balances, equipment specifications, and process optimization strategies under diverse scenarios, thereby enhancing overall process understanding and operational robustness [5].

Despite the breadth of research focused on biomass conversion and the advancements in simulation technologies, significant challenges remain. A notable gap persists between academic investigations and the full industrial implementation of simulated biomass conversion systems. This disconnect is driven by factors such as: (1) the lack of systematic consolidation of modeling approaches across diverse applications; (2) difficulties in translating simulation results into scalable industrial processes; and (3) the need for more comprehensive techno-economic and environmental feasibility assessments in real-world contexts.

This chapter provides a comprehensive overview of simulation-based strategies for biomass conversion, with an emphasis on modeling methodologies, simulator capabilities, and case studies reported in recent literature. The subsequent section focuses on the main usual processes for biomass reuse, covering various biomass conversion routes, feedstock characteristics, process conditions, and products of the possible alternatives. Next, Sects. 3 and 4 report the main characteristics of using chemical process simulators in biomass conversion, including the complexity of accurately representing biochemical reactions and biomass properties, as well as addressing emerging trends and the future outlook of industrial process simulators as important tools for advancing sustainable biomass utilization in the context of global bioenergy production. Finally, the promising advances and applications of process simulators for biomass conversion as a strategy to optimize industrial processes are addressed in Sect. 5. Effective biochemical conversion strategies are vital for turning biomass into diverse energy products, such as electricity, heat, biomethane, synthesis gas (which can be used to create hydrocarbons, ammonia, and methanol), biogas, bio-oil, bioethanol, biodiesel, among others [6]. The main biomass conversion routes from organic wastes, including rich-lipid wastes, are categorized as thermochemical or biological processes [7–9]. Figure 1 illustrates the primary raw materials, main processes, and products formed through these processes, which are briefly described hereafter.

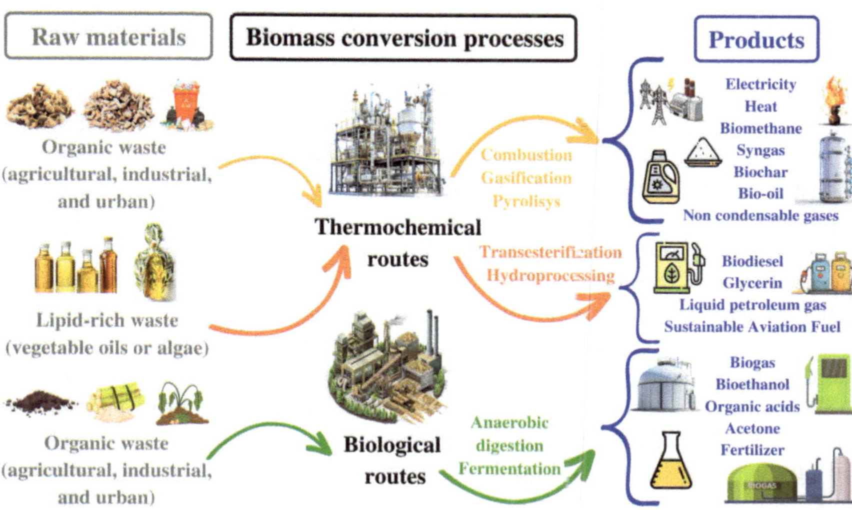

Fig. 1 Main technological biomass conversion processes of organic wastes and their featured products. (Source: organized by the authors with images created using Midjourney)

2 Usual Processes for Biomass Conversion and Reuse

2.1 Thermochemical Conversion of Organic Waste: Combustion, Gasification, and Pyrolysis

Thermochemical processes provide a flexible and efficient route for energy recovery from solid organic waste, which is often discarded. The choice of process should consider the energy purpose, raw material quality, system technical conditions, and environmental impacts.

The efficiency of the thermochemical process depends on several physicochemical properties of the raw material. Among these, the moisture content is highlighted, which should ideally be below 20% for gasification and pyrolysis, as high moisture reduces the energy yield. Regarding biomass composition, the proportions of essential elements, such as carbon, hydrogen, oxygen, sulfur, and nitrogen, affect energy release as well as pollutant formation [10]. Another factor to consider is ash content, since residues with high mineral content tend to cause fouling and reduce thermal efficiency. A high calorific value represents a greater amount of energy generated during combustion, which increases the energy yield. A smaller particle size facilitates reaction and thermal homogenization, thereby affecting heat transfer.

Combustion is the burning of biomass in the presence of oxygen to generate heat. The reaction occurs between hydrocarbons and oxygen and can be complete, with excess oxygen producing carbon dioxide (CO_2), or incomplete, with limited oxygen producing carbon monoxide (CO) or solid carbon in the form of ash. This is an ancient and straightforward energy conversion process that occurs at temperatures between 800 and 1000 °C, preferably with an excess of oxygen [10]. Typical products obtained are thermal energy, carbon dioxide (CO_2), water vapor (H_2O), and ash (inert solid residue). Main applications include direct heating, steam generation in biomass boilers, and electricity generation via the Rankine cycle.

Gasification is defined as a thermal process with limited oxygen, converting solid biomass into synthesis gas rich in CO and hydrogen (H_2) [11]. Notably, the temperature range of this reaction is typically between 800 and 1200 °C, similar to that of combustion. Main reactions include partial oxidation, where solid carbon and oxygen produce CO; steam reforming, where solid carbon and water produce CO and H_2; and water-gas shift reaction, where CO and water produce CO_2 and H_2 [10]. Main applications include electricity generation in engines and gas turbines, H_2 production, and the synthesis of liquid fuels via the Fischer-Tropsch process.

Finally, pyrolysis thermally decomposes biomass in the total absence of oxygen, generating liquid, solid, and gaseous fractions. At lower temperatures, between 500 and 800 °C, and in an inert atmosphere, typically nitrogen, the primary reactions involve the thermal cracking of macromolecules in the absence of oxygen [12]. The main products are bio-oil, a liquid fuel with medium energy density produced through fast pyrolysis, with yields of up to 70%; biochar, a carbon-rich solid obtained from slow pyrolysis, which is used as a soil conditioner; and volatile gases (CO, CO_2, CH_4, and H_2) [13].

2.2 Thermochemical Conversion of Lipid-Rich Waste: Vegetable Oils—Transesterification and Hydroprocessing

Transesterification is widely used to convert lipids from vegetable oils, waste oils, or algae biomass rich in triglycerides into methyl esters, known as biodiesel, with glycerol as a byproduct. The efficiency and kinetics of this reaction can be enhanced under controlled thermal conditions, which classifies this as a thermochemical process [14]. The quality and composition of lipids determine the rate of transesterification conversion. Lipid composition is based on triglycerides, and the proportion of fatty acids, degree of unsaturation, and presence of impurities such as free fatty acids and water influence the reaction. A high free fatty acid fraction may require pretreatment or acid-catalyzed processes since acids can react with base catalysts to form soaps, hindering product separation [15]. As with organic residues, moisture must be controlled, ideally kept below 0.5% to avoid undesired reactions. The oil source should be considered, as virgin seed or algae oils, and waste oils have varied compositions. Those with a higher triglyceride content are preferable, and waste oils require purification.

Transesterification describes the catalyzed reaction between triglycerides and an alcohol, usually methanol, to yield esters (the chemical foundation of biodiesel) as well as glycerol. This process consists of three successive reactions, each converting one acyl group of the triglyceride. Alkaline catalysts, including sodium hydroxide (NaOH) or potassium hydroxide (KOH), are frequently utilized on account of their high activity at moderate conditions (60–65 °C) and short reaction times (1–2 h). For high free fatty acid content, acid catalysts may be necessary, requiring higher temperatures and longer times [16]. A balance between methanol and triglyceride amounts is necessary, typically a 6:1 methanol-to-oil molar ratio. Excess methanol shifts the equilibrium toward biodiesel production, but it also increases costs and poses challenges to methanol recovery [17]. Biodiesel (methyl ester) is a renewable fuel compatible with conventional diesel engines, characterized by good viscosity, a high cetane number, and adjustable combustion properties that can be tailored to specific fatty acid profiles. It can be used in pure form or as an additive, reducing pollutant emissions and improving energy sustainability. Glycerol, a valuable byproduct, has broad applications in cosmetics, pharmaceuticals, and food industries, contributing to biodiesel's economic viability [18].

Transesterification provides an efficient route for converting lipids into biodiesel, positioning itself as a sustainable alternative to fossil fuels. Its feasibility depends on reaction conditions, feedstock quality, and the management of byproducts. Hydroprocessing routes are widely applied in petroleum refining to remove heteroatoms such as sulfur, nitrogen, and oxygen. More recently, these routes have been extended to biomass-based feeds [19]. Typically, triglyceride-rich feedstocks are converted into hydrocarbons using hydrogen at elevated temperatures and pressures in the presence of a heterogeneous catalyst. Coprocessing approaches, where biomass is blended with gasoils at various concentrations, have also been explored [19]. Unlike transesterification, hydrodeoxygenation (HDO) tolerates the presence of free fatty acids, which are not detrimental to the process itself. However, feedstock contaminants, such as metals and phospholipids, can be harmful to the catalyst [20]. Additionally, when the feedstock has low sulfur content, compounds may be added to maintain the catalyst in its sulfided form [21]. Triglycerides suitable for hydrodeoxygenation can be sourced from various origins, including virgin or waste vegetable oils, animal fats, and algal lipids [19]. Hydrodeoxygenation can also be combined with pyrolysis to process, although its contaminants, such as ashes and water content, may introduce catalyst deactivation [22]. The most accepted reaction mechanism involves saturation of double bonds, cleavage of the glycerol backbone, and three major deoxygenation pathways: hydrodeoxygenation (HDO), decarboxylation (DCX), and decarbonylation (DCN) [23]. While HDO produces hydrocarbons with even carbon numbers, DCX and DCN result in oddnumbered chains. Additional reactions such as methanation, cracking, cyclization, and isomerization may occur depending on the catalyst type and reaction conditions [23]. This is a more complex process than combustion, requiring controlled conditions of high pressure (typically 30–80 bar) and temperature (generally between 320 and 410 °C), in addition to the use of specific catalysts (such as MoS2 or WS2 promoted by Ni and supported on γ-Al_2O_3) [24]. The main products obtained are renewable hydrocarbons, including renewable diesel (HVO—Hydrotreated Vegetable Oil or "Green Diesel"), renewable jet fuel (hydrocarbons in the kerosene distillation range, suitable for use as sustainable aviation fuel (SAF)), renewable naphtha (lighter hydrocarbons, in the gasoline range), renewable LPG (propane, formed from the glycerol group present in the oils). Even though hydrodeoxygenation requires more severe conditions than transesterification, it offers the advantage of producing hydrocarbon fuels with greater chemical similarity to petroleum-derived counterparts.

2.3 *Biological Conversion of Organic Waste: Anaerobic Digestion and Fermentation*

Biological energy conversion processes represent a widely used and sustainable route for utilizing agroindustrial, urban, and lignocellulosic organic wastes. Based on microbial metabolic activity, these processes degrade complex organic matter into gaseous biofuels such as methane (CH_4) and H_2 [25]. Key biological mechanisms include anaerobic digestion, an established industrial process for biogas production in closed reactors, and fermentation, which occurs as an intermediate step in digestion or independently in optimized systems for H_2 generation. The main stages, operating conditions, products, and challenges of these biotechnological routes are detailed below.

Anaerobic digestion is a microbiological process that converts organic residues into biogas through a sequence of metabolic stages in an oxygen-free environment. Fermentation occurs in initial (hydrolytic and acidogenic) phases, while methane forms in the final methanogenesis stages. Four interdependent steps occur: hydrolysis, where macromolecules (polysaccharides, proteins, lipids) degrade into monomers (sugars, amino acids, fatty acids) by extracellular enzymes; acidogenesis (fermentation), where monomers are converted into volatile acids (acetate, propionate, butyrate), H_2, CO_2, NH_3, and H_2S by bacteria like Clostridium and Lactobacillus. Next is acetogenesis, converting volatile acids into acetate, H_2, and CO_2 by syntrophic bacteria. Finally, methanogenesis occurs, where methanogenic archaea (acetotrophic and hydrogenotrophic) convert acetate, H_2, and CO_2 into CH_4 and H_2O [26]. The main and most valuable product is biogas, a gaseous mixture primarily composed of CH_4 and CO_2, with small amounts of H_2, hydrogen sulfide (H_2S), and water vapor. As a versatile energy source, biogas can be used directly for heat and electricity or purified into biomethane, a renewable substitute for natural gas. Beyond energy use, biogas production mitigates greenhouse gases by preventing uncontrolled methane emissions from landfills and manure, and reduces pollutant loads in effluents and solid wastes [27]. Another product of anaerobic digestion is digestate, a nutrient-rich liquid residue containing nitrogen, phosphorus, and potassium, which can be used as a biofertilizer.

Fermentation is a metabolic process (molecular breakdown) through which organisms, predominantly microorganisms such as bacteria, and yeasts, and under certain conditions, even cells of higher organisms, obtain energy from the degradation of complex organic compounds, typically carbohydrates (sugars or starches), in an anaerobic environment (in the absence or with a very limited supply of oxygen) [28]. This process releases energy for the cell, but less efficiently than aerobic respiration. Alcoholic fermentation is the best-known example, where yeasts (such as Saccharomyces cerevisiae) convert sugars into ethanol, widely used as a biofuel, and carbon dioxide (CO_2) [29]. Other examples include the production of butanol and biogas (comprising methane and CO_2) through anaerobic fermentation [30].

Fermentation can generate a wide range of platform chemicals, including organic acids (such as lactic acid, citric acid, and acetic acid), solvents (such as acetone and butanol), and bioplastics (such as polylactic acid, or PLA) [31]. In some cases, the main objective of biomass fermentation is to produce protein-rich microbial biomass that can be used as animal feed or even for human consumption (e.g., biomass fermentation for alternative food sources and alcoholic drinks) [28]. After fermentation, the products of interest are separated from the fermented broth and purified to obtain the desired final product [32].

Anaerobic digestion and fermentation represent well-established routes for producing biogas, hydrogen, and valuable products from organic residues, offering significant environmental and energy benefits. Process efficiency depends on the proper control of operational parameters, substrate composition, and microbial management.

3 Chemical Process Simulators Characteristics

Chemical process simulators are indispensable tools for advancing and optimizing chemical processes on an industrial scale, being essential for the contemporary chemical industry, which also encompasses the biochemical industry linked to biomass processing. In general, a chemical process simulator consists of software or a set of programming scripts with a structure that enables the organization and correlation of mathematical models. This structure offers several advantages, particularly in solving models by applying specialized algorithms to systems of algebraic-differential equations, thereby presenting their solution.

In this scenario, a strong correlation is observed between the development of process simulators and the evolution of industry, especially with the emergence of computers and industrial automation. The progress of simulators as tools for chemical engineers, particularly those involved in process modeling and simulation, has been driven by the advancements of Industry 4.0. This includes the incorporation of features for managing large volumes of data (Big Data), such as neural networks, machine learning, and deep learning, in addition to Internet of Things (IoT) interfaces and cloud data storage. Classical and modern literature provide an in-depth examination of the characteristics and applications of various process simulators [33–37], which present a classification of process simulators, their historical context, architecture, examples, and their main applications. Despite distinct purposes and functionalities, advancements in the development of process simulators have enabled the identification of common factors. Process simulation tools imply a categorization and architecture that can be applied to various processes, adapting to the general purposes of simulators [35]. It is observed that process simulators must incorporate a database structure for components and their properties, thermodynamic solvers for calculating properties and phase equilibrium, flowsheet interfaces that allow the definition of process topology and their relationships, numerical method solvers capable of solving systems of algebraic-differential equations (for both static and dynamic processes), interactive result presentation interfaces, and global solvers to ensure the convergence of the entire process. To facilitate the understanding of this architecture, Fig. 2 illustrates the main stages and phases contained in chemical process simulators.

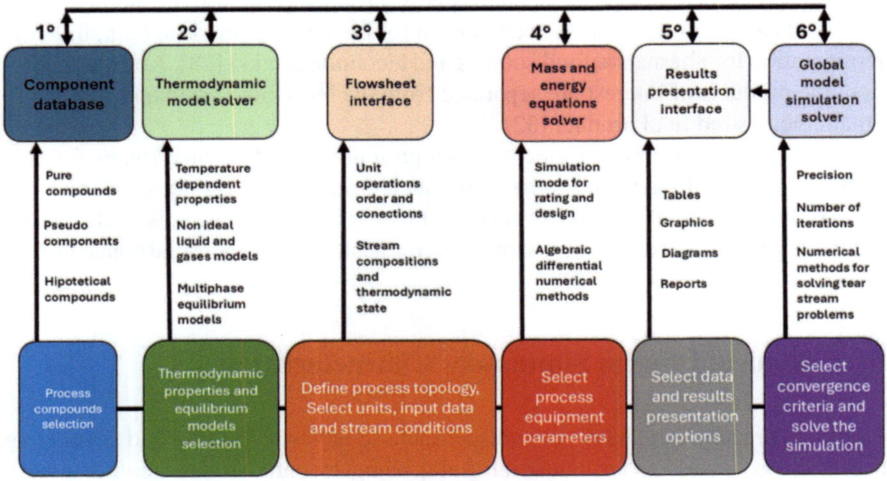

Fig. 2 General architecture of chemical process simulators. (Adapted from Turton et al. [35])

The presented architecture demonstrates that these stages are common in chemical process simulation in general; however, each simulator may incorporate additional supplements or functions according to its specific purpose. It is fundamental to emphasize that the distinction in the simulator's architecture for static and dynamic simulation purposes lies in the existence and necessity of solvers and numerical methods capable of solving specific models. Furthermore, presenting results for evaluating behavior over time requires graphical interfaces in dynamic simulators, as tabular visualization would be inefficient for extended periods of time. These are then considered the main intrinsic differences in the architecture of a chemical process simulator.

4 Process Simulators Applied to Biomass Conversion

In response to the global demand for sustainable energy, biomass conversion technologies have emerged as essential alternatives to fossil-based energy sources [9, 38]. However, the aforementioned processes involve complex reaction pathways that require meticulous study to ensure technological and economic viability, as the composition and properties of the substrate significantly impact the product yield and quality [39]. In this regard, process simulators play a critical role by offering powerful tools to model, analyze, and optimize these systems. In this chapter, the application of process simulation software to study the usual process for biomass conversion through thermochemical, biochemical, and physicochemical routes is discussed. An overview of all works discussed hereafter is presented in Table 1.

4.1 Combustion Conversion Processes

Recent simulation studies of biomass combustion have primarily focused on developing robust equipment models and estimating the composition of the resulting combustion products. The primary objectives of these efforts are to enable the optimization of the equipment design and operating conditions, enhance energy efficiency, and reduce pollutant.

Jia et al. [40] proposed a finite element simulation model to evaluate the influence of the excess air coefficient on the combustion performed in a rotary burner fueled by biomass wood pellets, using CFD software. The experimental data regarding the combustion temperature demonstrated a high level of agreement with the simulation results. The proposed model indicated that the middle part of the rotary burner presented the maximum recorded temperature. Furthermore, an increase in the excess air coefficient led to a decrease in the burner temperature and the amount of CO formed.

In this study focused on the combustion of sawdust biomass briquette in a chain boiler, Chen et al. [41]. identified the optimal parameters for maximizing both

Table 1 Framework of the investigated biomass conversion works: feedstocks, models, software and products

Process	Biomass consumed	Thermodynamic and/or fluid dynamic models (if applicable)	Simulator tools used	Products of interest in the study	Reference
Combustion	Wood pellets	k-ε for turbulence kinetic and stress equations	Ansys Fluent	Thermal energy and syngas	[41]
	Sawdust briquette	Porous media and k-ε for turbulence kinetic and stress equations	Ansys Fluent	Thermal energy and biogas	[42]
	Municipal solid waste	Finite rate/eddy dissipation and k-ε standard equations	FLIC and ANSYS Fluent	Thermal energy and syngas	[43]
Pyrolysis	Rice straw and sugarcane bagasse	Peng-Robinson	Aspen Plus	Biochar	[46]
	Synthetic lignocellulosic biomass	Peng-Robinson with Boston-Mathias modifications	Aspen Plus	Char, bio-oil, and syngas	[49]
	Food waste, cattle manure, and lignocellulosic biomass	Peng-Robinson	Aspen Plus	Bio-oil	[50]
Gasification	Pine pellets and chips derived from pine	Peng-Robinson with Boston-Mathias modifications	Aspen Plus	Syngas	[52]
	Rice husks	Peng-Robinson with Boston-Mathias modifications	Aspen Plus	Hydrogen-rich syngas	[54]
	Synthetic lignocellulosic biomass	Peng-Robinson with Boston-Mathias modifications	Aspen Plus	Syngas and tar	[55]

Table 1 (continued)

Process	Biomass consumed	Thermodynamic and/or fluid dynamic models (if applicable)	Simulator tools used	Products of interest in the study	Reference
Transesterification	Waste cooking oil	NRTL	Aspen Plus	Biodiesel	[58]
	Dry microalgae	NRTL and UNIFAC	Aspen Plus	Biodiesel	[59]
	Residual oils and fats	NRTL and UNIFAC	Aspen Plus	Biodiesel	[60]
Hydroprocessing	Palm oil	NRTL and SRK	Superpro Designer and VBA	Bio-jet fuel, green diesel and propane	[63]
	Macauba oil	Soave-Redlich-Kwong	Aspen Plus	Green diesel	[64]
	Soybean and tallow oil	Peng-Robinson	Unisim Design	Biokerosene and green diesel	[21]
	Palm oil components	Peng-Robinson and Redlich-Kwong	Aspen Plus	Green diesel	[65]
	Waste vegetable oil	Peng-Robinson	CHEMCAD	Green diesel	[66]
	Soybean oil	Peng-Robinson	Aspen Hysys	Green diesel and biokerosene	[67]
Aerobic digestion	Cow-manure	NRTL and Soave-Redlich-Kwong	Aspen Hysys	Biogas and hydrogen	[69]
	Food waste	NRTL	Aspen Plus	Biogas	[71]
	Cattle and poultry manure, slaughterhouse, and vegetable waste	RF, XGBoost, ANN, SVR and KNN	Python programming	Biogas	[72]
Fermentation	Sawmill waste	NRTL and ideal gas	ProSimPlus	Bioethanol	[74]
	Energy crops, forest residues, and agricultural residues	NRTL	Aspen Plus	Bioethanol and Electricity	[75]
	Corn cob	NRTL and SRK	Superpro Designer	Bioethanol	[76]

combustion efficiency and economic returns. Through the implementation of an ant colony algorithm, it was determined that the ideal conditions are an air-preheating temperature of 553.22 K, a fuel bed thickness of 0.16 m, and a particle diameter of 0.02 m, combined with a uniform-centered air supply system. This optimization was based on a validated model that integrated theories of computational fluid mechanics, heat transfer, thermodynamics, and chemical kinetics.

Huang et al. [42] explored the impact of biomass co-firing in a 600 t/day waste incinerator's grate furnace, focusing on temperature, pollutants, and denitrification. They simulated solid-phase combustion using the fluid dynamic incinerator code (FLIC) code and gas-phase combustion with ANSYS Fluent to simulate the gas-phase combustion [43]. The study observed that higher biomass blending ratios resulted in a gradual reduction in fuel moisture, an increased calorific value, and a corresponding rise in the maximum furnace chamber temperature.

4.2 *Pyrolysis Conversion Processes*

While experimental investigations are imperative for the precise design and subsequent optimization of pyrolyzers, the development of robust models to describe the pyrolysis reaction is essential to circumvent the costly and time-consuming nature of experimental investigations [44]. This section primarily focuses on simulating the pyrolysis reaction using both mechanistic models and data-driven models.

A study by Liu et al. [45] involved the development of a process model in Aspen Plus V.8.4 to simulate the thermal decomposition (pyrolysis) of rice straw and sugarcane bagasse. The feedstock, produced by Sande Corporation, was characterized through proximate and elemental analysis. The resulting data was then input into the simulator. The process biomass was represented by hemicellulose, cellulose, and lignin, modeled as $C_5H_8O_4$, $C_6H_{10}O_5$, and $C_{17}H_{17}O_5$, respectively, within the simulator. The representation of biomass by hemicellulose, cellulose, and lignin has also been performed in other simulation studies, such as those developed by Thu et al. [46] and Rosha et al. [47]. Liu et al. [45] employed a Gibbs reactor, which operates on the minimization of the total Gibbs free energy, to model the formation of biochar in their study. The proposed simulation model demonstrated a satisfactory capacity to estimate biochar yield within the temperature range of 300–500 °C. However, at elevated temperatures, the deviation from the experimental data attained values up to 8.93%. In the economic evaluation step, the estimated costs for biochar from rice straw were \$0.79/kg, while the cost for biochar from sugarcane bagasse was \$0.83/kg.

Shabaz et al. [48] simulated the slow pyrolysis process in Aspen Plus V.9 to investigate the influence of hemicellulose, cellulose, and lignin on the production of char, bio-oil, and syngas. The authors validated the model they developed by using data sourced from the scientific literature. A kinetic-free approach was employed, and the formation of tar, pressure drops, and heat losses was neglected to simplify the simulated models. The product formation was simulated using a yield reactor and a Gibbs free energy minimization reactor. These reactors were responsible for converting the biomass into conventional components based on the ultimate analysis of the biomass and calculating the chemical composition and phase equilibrium of the product, respectively. Under the given conditions of this study, lignin pyrolysis has been demonstrated to be a more advantageous process compared to the pyrolysis of cellulose and hemicellulose.

In their study, Ullah et al. [49] first developed a predictive model for bio-oil yield from biomass pyrolysis using a cascade forward neural network (CFNN). To optimize the process, they integrated this data-driven model with the artificial bee colony method, creating a system to find the input parameters that would maximize the bio-oil output. The CFNN model proved to be highly effective, demonstrating a correlation coefficient of 0.95. As a final step, the optimal conditions identified through this metaheuristic optimization were validated using a mechanistic simulation built in Aspen Plus V.10 software. The employed methodology for the simulation was similar to the methodology described previously in the work of Shabaz et al. [48]. The CFNN model achieved an average error of 5% when compared to the simulation model.

4.3 *Gasification Conversion Processes*

The process of biomass gasification has emerged as a subject of significant interest, given its potential application as a thermal degradation method that can contribute to the development of sustainable energy. However, accurately simulating biomass gasification poses significant challenges due to its reliance on reaction kinetics, reactor geometry, processing methodology, and operating conditions [50]. This section offers a concise overview of recent simulation methodologies for the biomass gasification process.

Using the Aspen Plus platform, Puig-Gamero et al. [51] recently conducted a simulation of a biomass gasification process occurring in a bubbling fluidized bed reactor. The equipment was divided in three main sections, where the pyrolysis, combustion, and reduction processes were evaluated. The pyrolysis zone was simulated using an external Microsoft Excel subroutine developed by [52]. This subroutine describes the yield and elemental composition of pyrolysis products. The gasifier's combustion and reduction reactions were represented by sets of kinetic equations, six for the former and seven for the latter, and kinetic parameters obtained from the literature. The proposed simulation was validated using experimental data obtained using an autothermal pilot-scale bubbling fluidized bed reactor. The revised model reliably predicted gasifier performance over various operating conditions, such as changes in temperature and equivalence ratio. Among the analyzed gases, H_2 exhibited the lowest prediction accuracy, with a tendency to be overestimated, reaching a maximum absolute error of 4.4%. Additionally, the model's predictive accuracy decreased for equivalence ratio values of 0.30 or higher.

Kombe et al. [53] developed a simulation of rice husk air gasification using Aspen Plus V.10 to investigate a process for producing hydrogen-rich syngas. Initially, the dried biomass was decomposed into its conventional components, as determined by ultimate and proximate analysis. This approach is analogous to the methodology previously discussed in the section on biomass pyrolysis. In this study, air was used as the gasifying agent, and oxidation and reduction reactions were simulated using RGibbs reactors. These reactors employ Gibbs free energy minimization to calculate the final syngas composition. A response surface methodology was employed to study the effects of temperature and equivalence ratio, the study identified that optimal process performance occurred at temperatures of 820–1090 °C as well as equivalence ratios of 0.06–0.10.

Abdul Azeez et al. [54] developed three simulation models for the biomass gasification process in Aspen Plus V. 8.8. The initial two models were developed using equilibrium and kinetic approaches. However, the third model incorporated reaction kinetics, gasifier geometry, and bed hydrodynamics. Models 1 and 2 employed methodologies analogous to those utilized in the studies conducted by Kombe et al. [53] and Puig-Gamero et al. [51], respectively. In Model 3, the same kinetic model that was employed in Model 2 was utilized. However, the incorporation of the reactor geometry, configuration, and bed hydrodynamic parameters was enabled by the Aspen Plus Fluidbed block. The model 3 demonstrated the most precise results, as indicated by the lowest values for mean error, root mean square error, mean absolute error, and mean absolute percentage error. Therefore, the model 3 provided the most accurate representation of the gasification process. Subsequent simulation experiments with model 3 demonstrated that elevated temperatures and reduced equivalence ratios enhanced the gasification performance.

4.4 *Transesterification and Vegetable Oil Oydroprocessing Conversion Processes*

Biodiesel production is a complex process that necessitates a structured approach to design and optimization [55]. Addressing this complexity effectively requires a comprehensive perspective that incorporates process design, quantitative evaluation, and optimization to improve overall system performance [56].

Liu et al. [57] evaluated the economic feasibility of waste cooking oil biodiesel production using the Life Cycle Cost method. In this study, the transesterification reaction was conducted with methanol and H_2SO_4 as the catalyst. The pretreatment of feedstock, production of biodiesel, and purification of the product were simulated using Aspen Plus. A sensitivity analysis in Aspen Plus was used to optimize the methanol and biodiesel rectification towers by varying the number of stages, feed stage, and reflux ratio. Economically, the biodiesel produced from waste cooking oil was found to be 65.28% more costly than diesel, a difference largely due to the required pretreatment of the feedstock to remove free fatty acids that interfere with the transesterification process.

A simulation and multi-objective optimization of a dry microalgae-to-biodiesel plant was conducted by Ahmed et al. [58]. The modeled process employed methanol as the alcohol and HCl as the catalyst for the conversion. To represent microalgal lipids and biodiesel in the simulation, the researchers used a pseudo-component model. Microalgal lipids, for instance, were modeled as a single triglyceride molecule comprised of representative fatty acids (palmitic, stearic, oleic, linoleic, and linolenic). This model was then used in a multi-objective optimization that aimed to improve the plant's overall performance by considering economic and environmental factors as constraints, targeting reductions in total annualized cost, organic waste, and CO_2 emissions. Additionally, proteins were represented by L-phenylalanine, while carbohydrates were modeled as sucrose. The process simulation was conducted using Aspen Plus V11, with the NRTL model selected for thermodynamic equilibrium. The simultaneous optimization of all objectives resulted in reductions in total annual cost (13.1%), organic waste generation (55%), and CO_2 emissions (41%).

In their study, Albuquerque et al. [59] proposed three solid acid-catalyzed processes for converting residual oils and fats. These processes were modeled in Aspen Plus V8.8, using reactive separation techniques. Triolein, oleic acid, methanol, methyl oleate, water, and glycerol were selected as representative components involved in the hydrolysis, esterification, and transesterification reactions. The case studies focused on simultaneous esterification, transesterification, and methanol separation through catalytic distillation, catalytic absorption, and hydroesterification with catalytic distillation. The techno-economic and environmental performances of these case studies were evaluated. While all processes were technically feasible, the optimized forms of the first two case studies were found to be more economical and environmentally friendly. The employment of simulations of vegetable oil hydroprocessing for the production of green diesel from triglyceride and fatty acid-based feedstocks performs a pivotal role in project development, allow comparisons of feeds, analysis of uncertainties, process sensitivity analysis, parameter optimization and key insights of the integrated processes [60–62]. These models are particularly valuable for economic optimization and estimating production costs. In the study by Martinez-Hernandez et al. [63], the authors performed a simulation and techno-economic analysis of a palm oil HDO plant to produce green diesel and bio-jet fuel, including steps of isomerization and cracking. The work applied the results from a proprietary catalyst in a yield reactor using SuperPro Designer simulator linked VBA to evaluate fuel constraints, sensitivity and uncertainty analyses. Their results indicate that the plant capacity was a determinant of the minimum selling price of fuel, and the process dedicated to green diesel is more likely to be economically viable. Besides, vegetable oil cost was found to be the major component of total annualized cost, followed by the annualized capital cost. A HDO simulation integrated with technoeconomic, envinronmental and geospatial analysis was performed by Salvador et al. [64]. They investigated the use of macauba oil in green diesel production in a study case in Brazil, considering different scenarios and production scales. The authors applied Aspen Plus V10 for the plant simulation considering Soave-Redlich-Kwong as thermodynamic package, from macauba beans to the refined fuel. Additionally, RenovaCalc was employed for the environmental evaluation and QGIS for geospatial analysis. Process feasibility, especially in smaller production scales, was found to be highly dependent on plant coproducts. At higher scale, the green diesel minimum was found to be comparable to those of fossil diesel. Macauba-generated diesel presented a carbon intensity of 11.23 gCO_2/MJ diesel and the highest potential for large scale plants was determined to be in Northeast region of Brazil. Seibel et al. [21] detailed patents related to the integrated HDO process and simulated hydrocarbon production from available data. The authors employed Unisim Design v.R491 using Peng-Robinson thermodynamic package to study soybean and tallow oil as feeds, which are relevant as biomass feeds for renewable fuels. The results achieved green diesel and biojetfuel according to specifications at 275 kg/h and 295 kg/h, respectively, from a 1000 kg/h feed and a 50 kg/h hydrogen consumption. The feed blend 75%wt soybean oil and 25%wt animal tallow was found to have lower hydrogen income while producing more biojetfuel, when compared to other blends. Plazas-Gonzáles et al. [65] simulated the hydrotreatment of palm oil components to produce green diesel in the presence of $NiMo/\gamma Al_2O_3$ using Aspen Plus V.7.3. The cracking of triglycerides, hydrogenation, decarboxylation, decarbonylation and fatty acid reduction reactions were modeled in an equilibrium reactor. The vegetable oil was simulated using triolein, tripalmitin, trilinolein and tristearin as components, while the Peng-Robinson thermodynamic model coupled with Redlich-Kwong was used to predict the gasliquid equilibrium of triglycerides at high pressures. Through a sensitivity analysis, the authors identified the following optimal reactor operating conditions: a pressure of 30–60 bar, a temperature between 300 and 400 °C, and a hydrogen-to-oil molar ratio of 20:1. In the context of these simulated conditions, a green diesel purity of 97% was successfully obtained. Wibowo et al. [66] used the CHEMCAD 7.1.2 simulation software to evaluate the green diesel production using waste vegetable oil as the feedstock. The triglycerides and fatty acids fraction of the oil were represented by triolein and oleic acid, respectively. The Peng-Robinson thermodynamic model was employed to estimate the phase equilibrium of both liquid and vapor phases. The isothermal equilibrium reactor model was used to simulate the decarbonylation, decarboxylation and hydrodeoxygenation reactions. The authors conducted a sensitivity analysis to evaluate the impact of reaction temperature and pressure on product yields. Pressure was varied from 20 to 120 bar, and temperature ranged from 250 to 1000 °C. The simulation indicated that the increase in pressure had a minimal effect on green diesel production, whereas the increase in temperature significantly decreased the yield and selectivity of the deoxygenation reactions. According to the simulations, the optimal operating conditions were identified as 250 °C, 20 bar, and a hydrogen:oil mole ratio of 1:12. Barbosa et al. [67] developed a process simulation model to evaluate the techno-economic challenges associated with the Brazilian green diesel production process. The authors of the study employed the Aspen HYSYS V.12 process simulator and also considered the production of biokerosene as a secondary product. Although soybean oil was initially selected as the feedstock due to its strong potential in the Brazilian market, the authors reported a lack of kinetic data regarding green diesel production with NiMo/Al_2O_3. Therefore, the kinetic data for producing green diesel from sunflower oil was used to calculate the yield of the decarboxylation, decarbonilation and hydrodeoxygenation reactions. The feedstock was represented by triolein, trilinolein, tripalmitin, trilinolenin, and tristearin, while the Peng-Robinson thermodynamic model was used for all simulation steps. In this case study, the green diesel production process was deemed not economically viable, primarily due to the high cost of the hydrogen compression system, vegetable oil, and electricity. However, the price of simulated green diesel production could be reduced by using existing petroleum refining infrastructure and with market prices could be achieved by reducing the price of soybean oil.

4.5 *Anaerobic Digestion Conversion Processes*

The modeling of the anaerobic digestion process is of fundamental importance for various purposes. It is essential for designing wastewater treatment systems and biogas power plants. Additionally, it is crucial for analyzing the sensitivity of plant performance to operational parameters and for monitoring and controlling the system's efficiency. Moreover, it is necessary for evaluating the feasibility of using new substrates with varying characteristics, biodegradability, and operating conditions [68].

In their work, Balaji et al. [69] modeled a combined process for renewable hydrogen production by linking two key stages: biomass anaerobic digestion and the dry reforming of the resulting biogas. The entire system is modeled and simulated using Aspen HYSYS, based on the simulation of an anaerobic digestor developed by Rajendran et al. [70]. Despite the fact that anaerobic digestion comprises four distinct stages, namely, hydrolysis, acidogenesis, acetogenesis, and methanogenesis, the hydrolysis step was not incorporated into the model due to challenges in specifying the biomass composition, as well as the difficulties in defining the physical properties of substances that are not included in existing databases. A plug-flow reactor model, a standard component within the HYSYS simulation suite, was utilized to simulate the dry reformation process. The model was configured to represent the conditions of an isothermal fixed-bed catalytic reactor. The NRTL and Soave-Redlich-Kwong thermodynamic models were used for the anaerobic digestion and dry reforming steps, respectively. The optimal conditions identified for this process consist of a biogas composition of 52% CH_4, 38% CO_2, and 10% H_2O, with a reactor temperature of 837.5 °C and a pressure of 101.3 kPa. Under these conditions, the process achieved a methane conversion rate of 89.9%, a CO_2 conversion of 99.99%, and a hydrogen selectivity of 1.21. The study's findings suggest that this integrated approach offers a technically viable and environmentally beneficial pathway for H_2 production.

Menacho et al. [71] utilized Aspen Plus to simulate the anaerobic digestion of food waste within a two-stage digester configuration. Their model was designed to evaluate process efficiency under conditions of escalating organic loading rates and hydraulic retention times. A central component of the investigation involved assessing the impact of three distinct fat concentrations, specifically 20%, 40%, and 60%. The simulation, which utilized the anaerobic digester model from Rajendran et al. [70] with a modified set of initial parameters, was validated against experimental data from the literature. This validation confirmed a high degree of accuracy, showing a maximum variance of only 2.5% relative to the experimental CH_4 concentration. The results from the Aspen Plus simulation further indicated that within an organic loading rate of 2–5 L/day, peak methane yields of 74.82% and 77.10% were achieved at fat concentrations of 40% and 60%, respectively.

In an effort to describe and predict the correlation between operational parameters and biogas production, Yildirin and Ozkaya [72] sought to model the relationship between key operational parameters and biogas yield by applying five different machine learning algorithms. Their work utilized a dataset from a full-scale anaerobic digestion plant in Balikesir, Turkey, with input variables including solids content, temperature, pH, alkalinity, and volatile fatty acids. A clear performance difference was observed: the Random Forest model provided the highest prediction accuracy ($R^2 = 0.9242$), whereas the K-Nearest Neighbor algorithm proved to be the least accurate ($R^2 = 0.8326$). Despite the inherent challenges associated with maintaining process stability during the anaerobic digestion process, due to the intricate nature of the system, Support Vector Machine algorithms achieved R^2 values of 0.8960, 0.8703, and 0.8655. Machine learning algorithms have demonstrated a remarkable capacity to accurately estimate biogas potential.

4.6 Fermentation Conversion Processes

Lignocellulosic biomass constitutes a promising feedstock for bioethanol production due to its widespread availability, low cost, potential for heat and power cogeneration using its non-fermentable components, and lack of direct competition with food resources [73].

Alio et al. [74] developed a simulation of an integrated biorefinery plant designed to convert various types of biomass, including sawdust and sawmill waste, into bioethanol. ProSimPlus 3.6, a French software package specifically designed for the simulation of multiphase processes, was employed to model the production of lignocellulosic ethanol. The NRTL model was the primary thermodynamic model employed in the simulation; however, the ideal gas model was utilized to simulate the combustion of gases. The study utilized the NREL databases to obtain data for components and properties not included in the primary software database. The bioethanol production process was modeled through several key steps—Organosolv pretreatment, enzymatic hydrolysis, fermentation, and distillation—using a framework derived from experimental data. The simulation demonstrated that the productivity of simulated ethanol equaled 351 L/ton of dry feedstock, processing 70,088 tons of biomass per year. The study concluded that the process simulations demonstrate the successful application of an ethanol/water pretreatment to a mixture of softwood species in a forest biorefinery.

Morales et al. [75] analyzed how seven types of lignocellulosic biomass, classified as energy crops, forest residues, or agricultural residues, affect the key balances (mass, carbon, water, and energy) of a biochemical bioethanol production process. Furthermore, their work involved a comparison of the yields obtained from each distinct unit operation. The process was simulated using Aspen Plus V.10 software, with the NRTL thermodynamic model. The proposed plant was divided into seven main sections, which corresponded to storage and chipping, pretreatment, enzymatic saccharification, the co-fermentation of both pentoses and hexoses, bioethanol recovery, cogeneration of heat and power, as well as wastewater treatment. The results of this study show that overall bioethanol yields vary significantly, with energy efficiency ranging from 19.0% to 29.0%, carbon efficiency from 27.3% to 46.2%, and overall yields from 19.0% to 31.0%. Switchgrass, which contains the highest concentrations of carbohydrates, has been demonstrated to produce the greatest bioethanol yields. In contrast, forest residues such as spruce have been shown to yield the lowest amounts of bioethanol. The composition of the feedstock has been demonstrated to significantly impact both water and carbon balances.

A process for producing fuel-grade bioethanol through corn cob fermentation was modeled [76] with Superpro Designer 8.5. Their simulation evaluated a production design composed of four main stages: pretreatment, hydrolysis, fermentation, and downstream processing. The simulation was executed in batch mode, as this configuration offers reduced capital expenditures and thereby offering the advantage of simpler instrumentation and control requirements than those needed for continuous or fed-batch modes of operation. Bioethanol is initially produced at a concentration of 10.85 wt% through direct fermentation. It is then purified to a concentration of 99.96%, meeting fuel-grade standards. The process yielded 0.35 g of ethanol per gram of corn cob, a figure that equates to a 44.3% yield relative to the available glucose and xylose content of the biomass. By-products such as lignin, carbon dioxide, and stillage can be utilized to produce phenolic compounds, serve as carbon sources for microbial growth, and function as liquid fertilizers, respectively.

5 Prospecting for Future Work and Challenges

This section outlines key directions for future research and development, emphasizing the need for innovative modeling approaches, enhanced computational tools, and seamless integration with emerging industrial technologies. The field of chemical process simulation has undergone remarkable advancements, transitioning from static, equilibrium-based analyses to dynamic, real-time predictive capabilities. Traditional simulators have proven indispensable in optimizing established petrochemical, pharmaceutical, and food industries [5]. However, as the global imperative for sustainability intensifies, driven by legislative frameworks and environmental concerns, the focus of process engineering is increasingly shifting towards renewable resources and green processes. Biomass conversion technologies, encompassing processes listed in Sect. 2 are central to this transition, offering pathways for the production of biofuels, green hydrogen, biogas, and biomethane, as well as a diverse array of green chemistry derivatives.

The inherent complexity of biomass feedstocks and their conversion mechanisms necessitates a significant evolution in existing simulation capabilities. Unlike homogeneous chemical feedstocks, biomass is highly heterogeneous, variable in composition, and often processed under extreme conditions, leading to intricate reaction networks and multiphase phenomena [6]. Then, current simulators often lack comprehensive models specific to biomass thermochemical and biochemical conversion pathways. Advances in the field should focus on developing and integrating robust, predictive conversion models for pyrolysis (e.g., torrefaction, fast pyrolysis, flash pyrolysis), gasification (e.g., air, steam, or oxygen gasification), and biodigestion (e.g., anaerobic digestion kinetics or models based on raw materials characterization). Then, future work should enhance:

Customizable Unit Operations: Enhancing the capability for users to define and integrate custom unit operation models that precisely represent novel biomass reactors (e.g., specialized fluidized bed reactors for gasification, innovative anaerobic digester designs). This could involve leveraging programming tools and scripting capabilities (e.g., Python, VBA, C++) within simulator environments, as already partially seen in some commercial and open-source packages (e.g., Aspen Custom Modeler, AVEVA and DWSIM).

Biochemical Pathway Integration: For biodigestion and fermentation, simulators require more granular control over complex metabolic pathways and interactions between microbial consortia or enzymatic contributions. Additionally, the inherently complex, non-linear, and often uncertain nature of biomass conversion processes makes them ideal candidates for the application of advanced computational intelligence techniques. These methods can significantly enhance model accuracy, predictive power, and optimization capabilities, moving beyond traditional deterministic approaches.

The application of industrial process simulators to biomass conversion represents a pivotal area for future research and development in chemical process engineering. While existing tools provide a foundational base, the unique complexities of biomass feedstocks and their conversion pathways demand significant advancements in modeling capabilities, the integration of cutting-edge computational intelligence, and the seamless adoption of Industry 5.0 paradigms. The integration of advanced computational paradigms, including artificial intelligence (AI) and the Industry 5.0 framework, presents both challenges and unparalleled opportunities for the next generation of industrial process simulators in the context of biomass conversion.

Therefore, the full realization of these advancements on process simulation will hinge on the widespread adoption and sophisticated utilization of digital twins interfaces, virtual replicas, powered by advanced simulators and real-time data, capable of enabling unprecedented levels of process understanding, real-time optimization, predictive maintenance, and robust decision-making across the entire lifecycle of biomass conversion facilities. Additionally, when combined with intuitive human-machine interfaces that incorporate virtual reality (VR) and augmented reality (AR), future process simulators will not only enhance operational efficiency but also foster a more collaborative, human-centric, and sustainable industrial ecosystem. The challenges are considerable, but the potential for transformative impact on green energy production and sustainable chemistry is immense, positioning advanced process simulation at the forefront of the bioeconomy's future.

6 Conclusions

In conclusion, industrial process simulators have become indispensable tools in the pursuit of efficient and sustainable biomass conversion technologies. Their ability to model complex systems, integrate thermodynamic and kinetic data, and perform optimization analyses enables researchers and engineers to explore a wide range of biochemical process configurations. Simulation environments such as Aspen Plus, Aspen Hysys, DWSIM, and SuperPro Designer offer robust frameworks for evaluating feedstock variability, reactive and separation systems, energy integration, and techno-economic aspects under realistic operating conditions.

Amid growing concerns about climate change and energy security, biomass emerges as a renewable and versatile feedstock for producing fuels, chemicals, and materials. However, effective utilization of biomass in the appropriate route requires careful consideration of environmental and economic parameters, which are often difficult to assess at early stages of development. By coupling simulation tools with life cycle assessment (LCA), techno-economic analysis, and experimental data, it becomes possible to generate models that support more sustainable and informed decision-making throughout the design and scale-up phases.

While simulation alone cannot replace experimental validation or pilot-scale trials, its strategic use accelerates innovation, reduces development costs, and improves the likelihood of successful industrial application. As global demand for renewable energy and biobased products continues to rise, industrial process simulation will likely remain a cornerstone in designing flexible, scalable, and environmentally responsible biomass conversion systems.

Looking ahead, the integration of advanced modeling approaches—such as multi-objective optimization, artificial intelligence, and dynamic simulation—offers promising avenues for the continued evolution of process simulators in biomass conversion processes. Furthermore, incorporating modular simulation environments and digital twins can enhance real-time monitoring, process control, and adaptive decision-making in industrial settings.

Acknowledgements The authors gratefully acknowledge financial support from the National Council for Scientific and Technological Development—CNPq (Brazil), processes 307958/2021-3 and 306085/2024-0, and the Coordenação de Aperfeiçoamento de Pessoal de Nível Superior—Brasil (CAPES), process 88887.953358/2024-00 and Finance Code 001.

References

1. Dobre, T.G., Marcano, J.G.S.: Chemical Engineering: Modelling, Simulation and Similitude. Wiley-VCH, Weinheim (2007)
2. Show, P.-L., Vo, D.-V.N., Ponnusamy, S.K., Rajendran Alta, S., Kumar, G.: Advanced thermochemical and biochemical processes for biomass transformation to biofuels and biochemicals. Biomass Convers. Biorefinery. (2024). https://doi.org/10.1007/s13399-024-05794-w
3. Safarian, S., Unnþórsson, R., Richter, C.: A review of biomass gasification modelling. Renew. Sust. Energ. Rev. **110**, 378–391 (2019). https://doi.org/10.1016/j.rser.2019.05.003
4. Kargbo, H., Harris, J.S., Phan, A.N.: "Drop-in" fuel production from biomass: critical review on techno-economic feasibility and sustainability. Renew. Sust. Energ. Rev. **135**, 110168 (2021). https://doi.org/10.1016/j.rser.2020.110168
5. Chaves, I.D.G., López, J.R.G., Zapata, J.L.G., Robayo, A.L., Niño, G.R.: Process Analysis and Simulation in Chemical Engineering. Springer, Berlin (2019)
6. Junior, O.V., Cordeiro, P.H.Y., da Cruz Ferrari, A.M., da Silva, L.L., Bonfim-Rocha, L.: Nanomaterials for catalytic conversion of biomass. In: Biomass Conversion Through Nanomaterials, pp. 151–183. Elsevier (2025)
7. Marafon, A.C., Santiago, A.D., Amaral, A.F.C., Bierhals, A.N., Paiva, H.L., Guimarães, V.S.: Uso da Biomassa para a Geração de Energia (In Portuguese). EMBRAPA. (2016) ISSN 1678-1953
8. Al-Mawali, K.S., Osman, A.I., Al-Muhtaseb, A.'a.H., Mehta, N., Jamil, F., Mjalli, F., Vakili-Nezhaad, G.R., Rooney, D.W.: Life cycle assessment of biodiesel production utilising waste date seed oil and a novel magnetic catalyst: a circular bioeconomy approach. Renew. Energy. **170**, 832–846 (2021). https://doi.org/10.1016/j.renene.2021.02.027
9. Osman, A.I., Mehta, N., Elgarahy, A.M., Al-Hinai, A., Al-Muhtaseb, A.'a.H., Rooney, D.W.: Conversion of biomass to biofuels and life cycle assessment: a review. Abstr. Environ. Chem. Lett. **19**(6), 4075–4118 (2021). https://doi.org/10.1007/s10311-021-01273-0
10. Basu, P.: Combustion and Gasification in Fluidized Beds. CRC Press/Taylor & Francis, Boca Raton (2006)
11. Belgiorno, V., De Feo, G., Rocca, C.D., Napoli, R.M.A.: Energy from gasification of solid wastes. Waste Manag. **23**(1), 1–15 (2003). https://doi.org/10.1016/S0956-053X(02)00149-6
12. Cortez, L.A.B., Lora, E.E.S., Gomez, E.O.: Biomassa para energia. UNICAMP (In Portuguese), Campinas, SP (2008)
13. Ogada, T., Werther, J.: Combustion characteristics of wet sludge in a fluidized bed: release and combustion of the volatiles. Fuel. **75**(5), 617–626 (1996). https://doi.org/10.1016/0016-2361(95)00280-4
14. Gao, Y., Mao, Y., Song, Z., Zhao, X., Sun, J., Wang, W., Chen, G., Chen, S.: Efficient generation of hydrogen by two-step thermochemical cycles: successive thermal reduction and water splitting reactions using equal-power microwave irradiation and a high entropy material. Appl. Energy. **279**, 115777 (2020). https://doi.org/10.1016/j.apenergy.2020.115777
15. Geris, R., dos Santos, N.A.C., Amaral, B.A., de Souza Maia, I., Castro, V.D., Carvalho, J.R.M.: Biodiesel de soja: reação de transesterificação para aulas práticas de química orgânica. Química Nova. **30**(5), 1369–1373 (2007). https://doi.org/10.1590/S0100-40422007000500053
16. Meher, L., Vidyasagar, D., Naik, S.: Technical aspects of biodiesel production by transesterification—a review. Renew. Sustain. Energy Rev. **10**(3), 248–268 (2006). https://doi.org/10.1016/j.rser.2004.09.002
17. Gomes, M.C.S.: Estudo da produção de biodiesel por transesterificação etílica e sua purificação utilizando processos com membranas (In portuguese). Doctoral Tesis, Universidade Estadual de Maringá, Maringá—PR, Brasil (2012)
18. Carvalho, G.G.M., Pinto, J.V.K.V.R.: Biodiesel: uma revisão de literatura do uso do glicerol residual. Instituto Federal do Espírito Santo, Vila Velha, ES (2023) (In portuguese)
19. Kubička, D., Tukač, V.: Hydrotreating of triglyceride-based feedstocks in refineries. In: Chemical Engineering for Renewables Conversion, pp. 141–194. Elsevier (2013)

20. Axens LLC: Vegan®—Technology for Premium Quality. Drop-in Biofuels From Renewable Oils & Fats. https://www.axens.net/expertise/renewable-fuels-bio-based-chemicals/saf-renewable-diesel (2021)
21. Seibel, J., Wancura, J.H.C., Mayer, F.D.: Process simulation and technology prospection to the hydrotreating of vegetable oils and animal fats. Energy Convers. Manag. **315**, 118811 (2024). https://doi.org/10.1016/j.enconman.2024.118811
22. Gollakota, A.R.K., Shu, C.-M., Sarangi, P.K., Shadangi, K.P., Rakshit, S., Kennedy, J.F., Gupta, V.K., Sharma, M.: Catalytic hydrodeoxygenation of bio-oil and model compounds—choice of catalysts and mechanisms. Renew. Sustain. Energy Rev. **187**, 113700 (2023). https://doi.org/10.1016/j.rser.2023.113700
23. Vázquez-Garrido, I., Guevara-Lara, A., López-Benítez, A.: Hydroprocessing of new and waste soybean oil for obtaining biodiesel: an operational conditions study. Chem. Eng. J. **452**, 139508 (2023). https://doi.org/10.1016/j.cej.2022.139508
24. Veriansyah, B., Han, J.Y., Kim, S.K., Hong, S.-A., Kim, Y.J., Lim, J.S., Shu, Y.-W., Oh, S.-G., Kim, J.: Production of renewable diesel by hydroprocessing of soybean oil: effect of catalysts. Fuel. **94**, 578–585 (2012). https://doi.org/10.1016/j.fuel.2011.10.057
25. Empresa de Pesquisa Energética – EPE: Hidrogênio e biomassa: oportunidades para produção e uso de hidrogênio em sistemas de bioenergia (In portuguese). Tecnical note DPG/SDB nº 02/2025, Brasília: EPE (2025)
26. Taiatele Junior, I.: Recuperação energética do biogás gerado a partir de água residuária de fecularias de mandioca: uma análise de viabilidade técnico-econômica e ambiental (In portuguese). Doctoral Tesis, Universidade Estadual de Londrina, Londrina – PR (2023)
27. ABiogás: ABiogás divulga novo potencial do biogás para o mercado brasileiro. In portuguese. https://abiogas.org.br/abiogas-divulga-novo-potencial-do-biogas-para-o-mercado-brasileiro-durante-forum-em-sao-paulo/ (2019)
28. Madigan, M.T., Bender, K.S., Buckley, D.H., Sattley, W.M., Stahl, D.A.: Brock Biology of Microorganisms. Pearson (2018)
29. Liu, Z., Wang, J., Nielsen, J.: Yeast synthetic biology advances biofuel production. Curr. Opin. Microbiol. **65**, 33–39 (2022). https://doi.org/10.1016/j.mib.2021.10.010
30. Lee, S.Y., Park, J.H., Jang, S.H., Nielsen, L.K., Kim, J., Jung, K.S.: Fermentative butanol production by clostridia. Abstr. Biotechnol. Bioeng. **101**(2), 209–228 (2008). https://doi.org/10.1002/bit.22003
31. Aui, A., Wang, Y., Mba-Wright, M.: Evaluating the economic feasibility of cellulosic ethanol: a meta-analysis of technoeconomic analysis studies. Renew. Sustain. Energy Rev. **145**, 111098 (2021). https://doi.org/10.1016/j.rser.2021.111098
32. Spalvins, K., Zihare, L., Dagnija, B.: Single cell protein production from waste biomass: comparison of various industrial by-products. Energy Procedia. **147**, 409–418 (2018). https://doi.org/10.1016/j.egypro.2018.07.111
33. Dimian, A.-C.: Integrated Design and Simulation of Chemical Processes. Elsevier (2003)
34. Foo, D.C.Y.: Chemical Engineering Process Simulation. IChemE, Elsevier (2019)
35. Turton, R., Bailie, R.C., Whiting, W.B., Shaeiwitz, J.A., Bhattacharyya, D.: Analysis, Synthesis and Design of Chemical Processes. Prentice Hall (2018)
36. Puigjaner, L., Ollero, P., de Prada, C., Jiménez, L.: Estratégias de projeto, simulação e otimização de processos químicos. Editorial Sinteis S. A, Madrid (2006)
37. Haydary, J.: Chemical Process Design and Simulation: Aspen Plus and Aspen HYSYS Applications, New Delhi, John Wiley & Sons, Inc (2019)
38. Pang, S.: Advances in thermochemical conversion of woody biomass to energy fuels and chemicals. Biotechnol. Adv. **37**(4), 589–597 (2019). https://doi.org/10.1016/j.biotechadv.2018.11.004
39. Poornima, S., Manikandan, S., Prakash, R., Deena, S.R., Subbaiya, R., Karmegam, N., Kim, W., Govarthanan, M.: Biofuel and biochemical production through biomass transformation using advanced thermochemical and biochemical processes—a review. Fuel. **372**, 132204 (2024). https://doi.org/10.1016/j.fuel.2024.132204

40. Jia, G., Li, L., Zhang, D.M.: Effect analysis on combustion and emission characteristics of a rotary burner fueled by biomass pellet fuel. J. Chem. **2020**, 1–12 (2020). https://doi.org/10.1155/2020/3618382
41. Chen, R., Yue, H.H., Yue, R., Ai, Y., Zheng, J.X.: Numerical simulation of combustion in a biomass briquette chain boiler. Biomass Convers. Biorefin. **11**(5), 1521–1536 (2021). https://doi.org/10.1007/s13399-019-00569-0
42. Huang, Z., Gong, D., Peng, C., Chen, J., Luo, J., Xu, Y., Yang, L.: Numerical simulation study on combustion of low calorific value waste blended with biomass. ACS Omega. **9**(12), 14297–14309 (2024). https://doi.org/10.1021/acsomega.3c10104
43. Yang, Y.B., Yamauchi, H., Nasserzadeh, V., Swithenbank, J.: Effects of fuel devolatilisation on the combustion of wood chips and incineration of simulated municipal solid wastes in a packed bed☆. Fuel. **82**(18), 2205–2221 (2003). https://doi.org/10.1016/S0016-2361(03)00145-5
44. Vuppaladadiyam, A.K., Varsha, S.S., Sikarwar, V.V.S., Ahmad, E., Pant, K.K., Murugavelh, S., Pandey, A., Bhattacharya, S., Sarmah, A., Leu, S.-Y.: A critical review on biomass pyrolysis: reaction mechanisms process modeling and potential challenges. J. Energy Inst. **108**, 101236 (2023). https://doi.org/10.1016/j.joei.2023.101236
45. Liu, Y., Yang, X., Zhang, J., Zhu, Z.: Process simulation of preparing biochar by biomass pyrolysis via aspen plus and its economic evaluation. Waste Biomass Valor. **13**(5), 2609–2622 (2022). https://doi.org/10.1007/s12649-021-01671-z
46. Thu, K., Reungpeerakul, T., Sangwichien, C.: Simulation of reaction kinetics and heat transfer effects on product yields from fast pyrolysis of oil palm empty fruit bunch biomass in fluidized bed reactor. BioEnergy Res. **13**(4), 1194–1204 (2020). https://doi.org/10.1007/s12155-020-10148-1
47. Rosha, P., Kumar, S., Ibrahim, H.: Sensitivity analysis of biomass pyrolysis for renewable fuel production using Aspen Plus. Energy. **247**, 123545 (2022). https://doi.org/10.1016/j.energy.2022.123545
48. Shahbaz, M., AlNouss, A., Parthasarathy, P., Abdelaal, A.H., Mackey, H., McKay, G., Al-Ansari, T.: Investigation of biomass components on the slow pyrolysis products yield using Aspen Plus for techno-economic analysis. Abstr. Biomass Convers. Biorefin. **12**(3), 669–681 (2022). https://doi.org/10.1007/s13399-020-01040-1
49. Ullah, Z., Khan, M., Naqvi, S.R., Khan, M.N.A., Farooq, W., Anjum, M.W., Yaqub, M.W., AlMohamadi, H., Almomani, F.: An integrated framework of data-driven metaheuristic and mechanistic modeling approach for biomass pyrolysis. Process Saf. Environ. Protect. **162**, 337–345 (2022). https://doi.org/10.1016/j.psep.2022.04.013
50. Yu, J., Smith, J.D.: Validation and application of a kinetic model for biomass gasification simulation and optimization in updraft gasifiers. Chem. Eng. Process. Process. Intensif. **125**, 214–226 (2018). https://doi.org/10.1016/j.cep.2018.02.003
51. Puig-Gamero, M., Pio, D.T., Tarelho, L.A.C., Sánchez, P., Sanchez-Silva, L.: Simulation of biomass gasification in bubbling fluidized bed reactor using Aspen Plus®. Energy Convers. Manag. **235**, 113981 (2021). https://doi.org/10.1016/j.enconman.2021.113981
52. Neves, D., Thunman, H., Matos, A., Tarelho, L., Gómez-Barea, A.: Characterization and prediction of biomass pyrolysis products. Prog. Energy Combust. Sci. **37**(5), 611–630 (2011). https://doi.org/10.1016/j.pecs.2011.01.001
53. Kombe, E.Y., Lang'at, N., Njogu, P., Malessa, R., Weber, C.-T., Njoka, F., Krause, U.: Process modeling and evaluation of optimal operating conditions for production of hydrogen-rich syngas from air gasification of rice husks using aspen plus and response surface methodology. Bioresour. Technol. **36**, 1127734 (2022). https://doi.org/10.1016/j.biortech.2022.127734
54. Abdul Azeez, K.T., Suraj, P., Muraleedharan, C., Arun, P.: Aspen Plus simulation of biomass gasification: a comprehensive model incorporating reaction kinetics hydrodynamics and tar production. Process Integr. Optim. Sustain. **7**(1–2), 255–268 (2023). https://doi.org/10.1007/s41660-022-00291-x

55. Nasir, N.F., Daud, W.R.W., Kamarudin, S.K., Yaakob, Z.: Process system engineering in biodiesel production: a review. Renew. Sustain. Energy Rev. **22**, 631–639 (2013). https://doi.org/10.1016/j.rser.2013.01.036
56. Pasha, M.K., Dai, L., Liu, D., Guo, M., Du, W.: An overview to process design simulation and sustainability evaluation of biodiesel production. Abstr. Biotechnol. Biofuels. **14**(1) (2021). https://doi.org/10.1186/s13068-021-01977-z
57. Liu, Y., Yang, X., Adamu, A., Zhu, Z.: Economic evaluation and production process simulation of biodiesel production from waste cooking oil. Curr. Res. Green Sustain. Chem. **4**, 100091 (2021). https://doi.org/10.1016/j.crgsc.2021.100091
58. Ahmed, M., Abdullah, A., Laskar, A., Patle, D.S., Vo, D.-V.N., Ahmad, Z.: Process simulation and stochastic multiobjective optimisation of homogeneously acid-catalysed microalgal in-situ biodiesel production considering economic and environmental criteria. Fuel. **327**, 125165 (2022). https://doi.org/10.1016/j.fuel.2022.125165
59. Albuquerque, A.A., Ng, F.T.T., Danielski, L., Stragevitch, L.: Reactive separation processes applied to biodiesel production from residual oils and fats: design optimization and techno-economic assessment of routes using solid catalysts. Energy. **240**, 122784 (2022). https://doi.org/10.1016/j.energy.2021.122784
60. Lucantonio, S., Di Giuliano, A., Rossi, L., Gallucci, K.: Green diesel production via deoxygenation process: a review. Energies. **16**(2), 844 (2023). https://doi.org/10.3390/en16020844
61. Silva, G.C.R., de Andrade, M.H.C.: Simulation of deoxygenation of vegetable oils for diesel-like fuel production in continuous reactor. Biomass Convers. Biorefin. **13**(3), 1843–1857 (2023). https://doi.org/10.1007/s13399-021-01441-w
62. Diniz, A.P.M.M., Sargeant, R., Millar, G.J.: Stochastic techno-economic analysis of the production of aviation biofuel from oilseeds. Biotechnol. Biofuels. **11**(1) (2018). https://doi.org/10.1186/s13068-018-1158-0
63. Martinez-Hernandez, E., Ramírez-Verduzco, L.F., Amezcua-Allieri, M.A., Aburto, J.: Process simulation and techno-economic analysis of bio-jet fuel and green diesel production—minimum selling prices. Chem. Eng. Res. Des. **146**, 60–70 (2019). https://doi.org/10.1016/j.cherd.2019.03.042
64. Salvador, G.V., Salim, V.M.M., Toniolo, F.S.: Sustainability assessment of a decentralized green diesel production in small-scale biorefineries. Abstr. Biofuels Bioprod. Biorefin. **16**(6), 1527–1550 (2022). https://doi.org/10.1002/bbb.2419
65. Plazas-González, M., Guerrero-Fajardo, C.A., Sodré, J.R.: Modelling and simulation of hydrotreating of palm oil components to obtain green diesel. J. Clean. Prod. **184**, 301–308 (2018). https://doi.org/10.1016/j.jclepro.2018.02.275
66. Wibowo, A.A., Mustain, A., Lusiani, C.E., Hartanto, D., Gintint, R.R.: Green diesel production from waste vegetable oil: a simulation study. AIP Conf. Proc. **2223**, 020008. https://ez48.periodicos.capes.gov.br/ (2020). https://doi.org/10.1063/5.0000925
67. Barbosa, I.V., Scapim, L.A., Cavalcante, R.M., Young, A.F.: Industrial production of green diesel in Brazil: process simulation and economic perspectives. Renew. Energy. **219**, 119591 (2023). https://doi.org/10.1016/j.renene.2023.119591
68. Ficara, E., Hassam, S., Allegrini, A., Leva, A., Malpei, F., Ferretti, G.: Anaerobic digestion models: a comparative study. IFAC Proc. Vol. **45**(2), 1052–1057 (2012). https://doi.org/10.3182/20120215-3-AT-3016.00186
69. Balaji, R.K., Rajan, K.P., Ragula, U.B.R.: Modeling & optimization of renewable hydrogen production from biomass via anaerobic digestion & dry reformation. Int. J. Hydrogen Energy. **45**, 18226–18240 (2020). https://doi.org/10.1016/j.ijhydene.2019.08.209
70. Rajendran, K., Kankanala, H.R., Lundin, M., Taherzadeh, M.J.: A novel process simulation model (PSM) for anaerobic digestion using Aspen Plus. Bioresour. Technol. **168**, 7–13 (2014). https://doi.org/10.1016/j.biortech.2014.01.051
71. Menacho, W.A., Mazid, A.M., Das, N.: Modelling and analysis for biogas production process simulation of food waste using Aspen Plus. Fuel. **309**, 122058 (2022). https://doi.org/10.1016/j.fuel.2021.122058

72. Yildirim, O., Ozkaya, B.: Prediction of biogas production of industrial scale anaerobic digestion plant by machine learning algorithms. Chemosphere. **335**, 138976 (2023). https://doi.org/10.1016/j.chemosphere.2023.138976
73. Quintero, J., Cardona, C.A.: Process simulation of fuel etanol production from lignocellulosics using Aspen Plus. Ind. Eng. Chem. Res. **50**, 6205–6212 (2011). https://doi.org/10.1021/ie101767x
74. Alio, M.A., Marcati, A., Pons, A., Vial, C.: Modeling and simulation of a sawdust mixture-based integrated biorefinery plant producing bioethanol. Bioresour. Technol. **325**, 124650 (2021). https://doi.org/10.1016/j.biortech.2020.124650
75. Morales, M., Arvesen, A., Cherubini, F.: Integrated process simulation for bioethanol production: effects of varying lignocellulosic feedstocks on technical performance. Bioresour. Technol. **328**, 124833 (2021). https://doi.org/10.1016/j.biortech.2021.124833
76. González-Contreras, M., Hernández-Escoto, H., Aguilar-Garnica, E.: A comprehensive analysis of bioethanol and ethyl lactate joint production in second-generation biorefinery: simulation techno-economic and profitability assessments. Bioresour. Technol. **385**, 129470 (2023). https://doi.org/10.1016/j.biortech.2023.129470

DFT Simulations on Reactivity, Solvent, and Nitrogenous Base Pair Interactivity of the Ru-(II)-η6-p-Cymene-Trifluorbenzylphosfone Complex

Guilherme Bonifácio Rosa, Alexandre Junio Camilo, Karen Wohnrath, and Sergio Ricardo de Lazaro

Abstract Cancer is one of the most ancient disease known to humanity and its treatment remains complex and difficult. In the 1960 decade, studies and applications of coordination complexes are very promising in cancer treatment and a wide class of complexes were developed. Among the complexes, Ru-complexes are notorious for chemical properties that mimic Fe atoms properties in human metabolism further present a wide possibility of functional ligands. Ru (II)-η6-p-cymene-trifluorbenzylphosfone complex (Ru-cyTPh) is compose by p-cymene ligand that stabilizes Ru oxidation states and charge transfer mechanism, and phosphine ligand that is widely studied by large range of modification in structure. The DFT/B3LYP of Ru-cyTPh investigated the molecular stability of the molecule and frontier orbital interaction with Guanine-Cytosine (G-C) and Adenine-Thymine (A-T) nitrogenous basis pair. Electronic characterization by Pearson theory and Koopman theorem, predicts that Ru-cyTPh interacts with G-C basis pair specifically with Guanine molecule which reduces Ru-complex.

Keywords Cancer · Guanine · Adenosine · DNA · Ruthenium · DFT · Solvent effect

1 Introduction

In modern times, with high technology and advanced industrialization, one of the many symptoms of societal evolution is the interconnection between planetary health and human health [1–5]. With a wide variety of body tissues and consequently, a variety of cells in the human body, the problem of cancer is very complex

G. B. Rosa · A. J. Camilo · K. Wohnrath · S. R. de Lazaro (✉)
State University of Ponta Grossa, Ponta Grossa, Paraná, Brazil
e-mail: srlazaro@uepg.br

C. A. Taft, S. R. d. Lazaro (eds.), *Innovative Materials for Energy and Sustainability*, Engineering Materials,
https://doi.org/10.1007/978-3-032-09149-9_5

and demands specific study and diagnosis for each type of tumor. The most commons cancers diagnosed are the endometrial, Gallbladder, Head and neck, Hodgkin lymphoma, kidney, leukemia, liver, lung, mesothelioma, esophageal, ovarian, pancreatic, penile, prostate, skin, stomach, testicular, thyroid, vaginal and vulvar cancers [6]. Each cancer has a specified treatment and a medical approach for the patient. This is the reason for cancer treatment, which is very complex and challenging [1, 7–11]. In 2022, the International Agency for Research on Cancer (IARC) registered a close of 20 million new cases and 9.7 million deaths from cancer [12]. From 1975 to 2021, the number of cases and consequent deaths increased. Although in the 1990s, the number of fatalities decreased because of the many treatments and scientific developments in health [13]. This is a consequence of decades of exhaustive medicine, chemistry, biology, and physical research [14–16].

Cancer is natural for all forms of life, humans, animals, or plants [17, 18]. This is because there are anomalies in DNA transcriptions, and all cell regenerations have the essential step of replication of DNA. This replication occurs on a nitrogenous basis that composes the DNA structure, which includes purines such as Adenine (A), Guanine (G), and pyrimidines Cytosine (C), and Thymine (T). The sequence of these nitrogenous bases forms a helix, and a posteriori combination with a second helix builds the double helix DNA structure. The double helix of DNA structure presents hydrogen bonds between Adenine-Thymine (A-T) and Guanine-Cytosine (G-C) base pairs [19, 20]. Each sequence in DNA presents a biological function, such as metabolic functionality, protein synthesis, tissue growth and regeneration, and the apoptosis process. The error in this sequence in DNA structure causes the commitment of the biological function of an organism.

The unbridled growth of cells forms great tissues in the body, which can develop into a tumor. Another example is the uncontrolled replication and growth of white blood cells in the blood, leukemia disease [21].

The first medical description of cancer dates to 1600 B.C. in the Edwin Smith Surgical Papyrus, and studies suggest that this manuscript is a copy of an ancient Egyptian text from around 2500 B.C., contemporary to the construction of the Giza pyramids [21]. In this ancient manuscript, the illness is described as "incurable". Cancer cases have been reported throughout human history. Hippocrates of Cos, around 300 B.C., associated the tumors with crab legs, creating the denominations of cancer and carcinoma, classifying them as benign and malignant tumors. Cases in the Roman period written by Aulus Cornelius Celsus [22] describe surgical removal of breast cancers from 25 B.C. to 50 B.C., while it is reported that Archigenes of Apameia, in Greek, rejected surgical treatment and used drugs and medicinal plants in treatment. The most famous report is attributed to Galen of Pergamum, who associated cancer with the excretions of bile in the liver, as impurity in the human body [23]. Both cases did not advance in their respective treatments. In the Byzantine Empire and Islamic medicine in the Middle East around the year 1000, the main technique was removing the cancer, followed by cauterization, without a cure for any patient. Nevertheless, such a technique was widely utilized in the middle-aged period [24].

Until the nineteenth century, medicine was largely archaic and invasive, relying heavily on surgical treatments. Most advances were based on empirical surgery and clinical observations [25]. It is registered as a surgical procedure for the removal of breast cancer with extreme surgery, which was made possible by the discovery of anesthesia [26]. The twentieth century was pivotal for medicine due to the significant advancements in science, particularly in physics and chemistry. This fact was fundamental in understanding why many medicinal treatments were created, improved, or refined for application. At the end of the nineteenth century, with no atomic or molecular models, a significant gap existed between medicine and chemistry. Organic synthetic chemistry at the end of the nineteenth century and the beginning of the twentieth century can be resumed in the application of dye in the textile industry [27, 28].

2 Treatment with Organic and Inorganic Species

In 1908, the first treatment with organic and inorganic species such as arsenic trioxide, salvarsan, melarsoprol, dimethyl arsenic acid, GSO, and ZIO-101 molecules (Fig. 1) [29, 30]. Chemotherapy was born in Germany in 1910. Paul Ehrlich wrote about the possibility of cancer treatment with an in-depth study of the chemical properties of pharmacologic compounds [27, 31–33]. The first significant advance was the in vitro isolation of tumoral cells from animals and subsequent characterization of cancer's metabolic and chemical behavior, as well as differentiation in relation to healthy cells.

Although the arsenic treatment has presented an increase in tumors in patients, the results were completely unexpected by the medical community. This is because

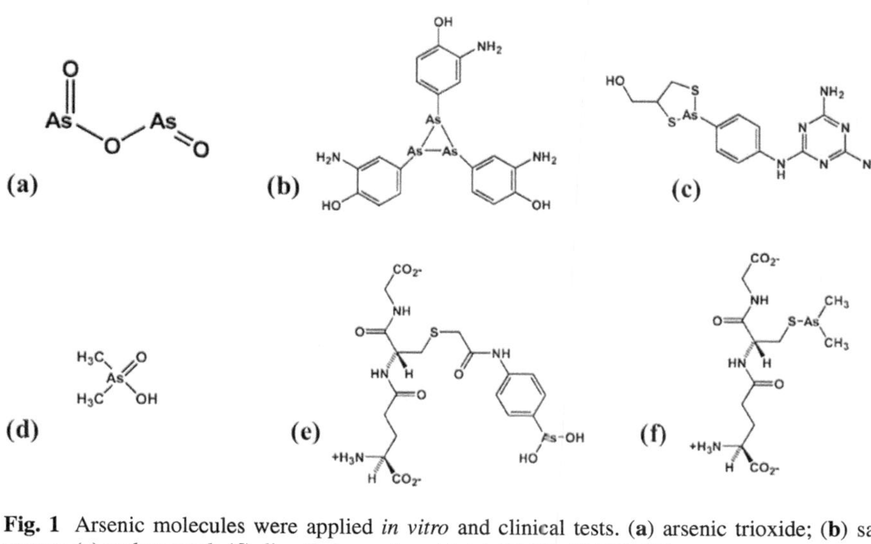

Fig. 1 Arsenic molecules were applied *in vitro* and clinical tests. (**a**) arsenic trioxide; (**b**) salvarsan; (**c**) melarsoprol; (**d**) dimethyl arsenic acid; (**e**) GSAO; (**f**) ZIO-101

Arsenic is toxic above 0.07 mg kg^{-1} in human metabolism [34, 35]. In 1930, the USA started cataloging chemical substances to research properties in the human body. From this research and the First World War period, the molecules investigated in patients were mustard gas analogues (Fig. 2). During the tests, the mustard gas analogues were promising. However, the high toxicity and very low selectivity caused the death of healthy cells. The toxicity of such compounds was the determining factor in discarding treatment with these molecules [36, 37].

In 1950, pediatric pathologist Sidney Faber, the father of chemotherapy, conducted tests at the Children's Hospital of Boston. The tests were carried out on patients in a terminal state. The primary cancer studied by Faber was Leukemia, which for a long time was considered an incurable disease [38, 39]. The primary issue with this cancer is its impact on mobility. While traditional cancer is well localized in solid tumors, Leukemia is a cancer of uncontrollable growth or out of metabolic control of the white cells present in blood, the leukocytes. The consequence is blood with a white color and a diluted aspect, making its treatment difficult even nowadays. Faber's research accelerated the evolution of cancer in patients through the application of folic acid (Fig. 3a). The consequence was innumerable deaths from Leukemia. From these results, the next step was to inhibit folic acid metabolism with antifolate analogues of Leukemia in patients (Fig. 3b). The role of folic acid in metabolism is essential to cellular growth, specifically in purine and pyrimidine synthesis, affecting DNA replication and protein synthesis [40]. This is the first time in history that humanity has applied chemical molecules to cancer treatment. Although long-term, the patients presented an improvement in health, after a 1- or 2-week period, Leukemia returned to manifest in a more aggressive state, causing the death of the patients [41–43]. Studies with antifolic compounds have shown the potential of using chemistry for cancer treatment. Furthermore, the 1960s marked the beginning of significant advancements in chemical compounds (Fig. 3c–f).

The significant advance in chemotherapy that occurred in the 1960s and 1970s was the development of metallic coordination complexes, specifically cis-$[PtCl_2(NH_3)_2]$, also widely known as cisplatin. Such a molecule became the most notorious compound discovered for cancer treatment in the twentieth century. Oher Pt complexes analogue to cisplatin was the $\{Pt[C_4H_6(CO_2)_2][NH_3]_2\}$ or Carboplatin [44–46]. The first tests of cisplatin and carboplatin were in terminal patients with testicular and ovarian cancers [47, 48]. Again, research and applications with Pt complexes showed the problem of the toxicity and low selectivity between healthy and cancer cells [49]. Although these Pt complexes were

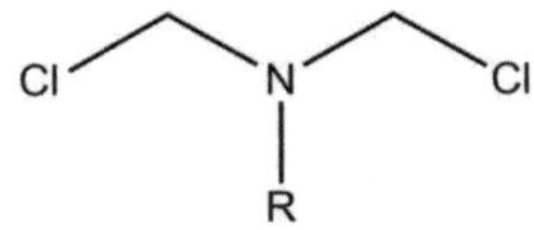

Fig. 2 Nitrogenous compounds are derived from mustard gas. The wide variety of compounds is centered on the R group

Fig. 3 Molecular structures of (**a**) folic acid and antifolic compounds as (**b**) aminopterin, (**c**) methotrexate, (**d**) pralatrexate, (**e**) raltritrexed, and (**f**) pametrexed, both compounds utilized by Sidney Faber in Leukemia treatment

fundamental to the scientific community's study, developing more metallic complexes with anti-cancer behavior was necessary (Fig. 4). For the 1970s, transition metals such as Ag, Pd, Au, Cu, Rh, Ir, Os, Re, Pt, and Ru complexes (Fig. 5) were developed for cancer treatment [50–54]. Transition metals are attractive because they offer a wide range of possibilities for geometries, ligands, and oxidation states. With modulation of these characteristics, it is possible to modulate the chemical properties to a specific application [55]. Among these transition metals mentioned, the Ru complex is very promising in cancer drug applications [56].

Ruthenium complexes are widely applied in metal alloys, solar cells, electronic devices, catalysts, and cancer drugs [55–63]. In fact, the Ru presents many applications for wide oxidation states, but the most stable and common is $Ru^{2+/3+}$ states. As is well known, the properties of the metallic center depend on the ligands in the molecule. Ligands control chemical behavior, such as stability of redox processes,

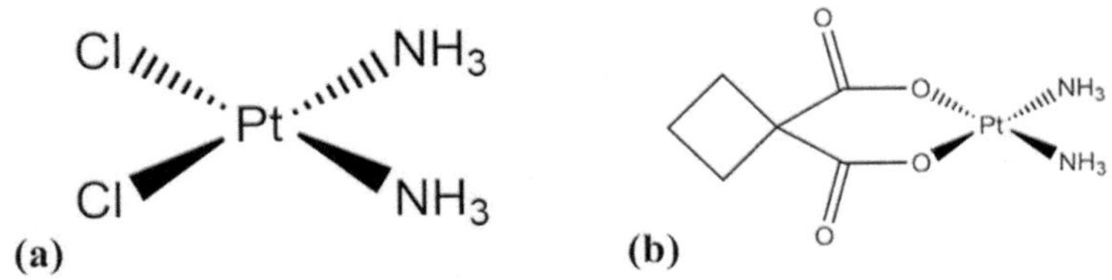

Fig. 4 Molecular structure of Pt complexes (**a**) cis-[PtCl2$(NH_3)_2$], the cisplatin, and (**b**) {Pt[$C_4H_6(CO_2)_2$][NH_3]$_2$}, the carboplatin

charge transfer, reactivity, and catalyst behavior, for example. In medicinal chemistry, specifically in oncology, Ru complexes are a modern and sophisticated alternative to cisplatin analogues [64–66] (Fig. 4). The first generation of Ru complexes applied was chloride-amin-Ru(III) $RuCl_3(NH_3)_3$ and [cis-$RuCl_2(NH_3)_4$]Cl and pair of isomers cis-$RuCl_2$(dimethyl-sulfoxide)$_4$ and trans-$RuCl_2$(dimethyl-sulfoxide)$_4$ (Fig. 5a–d) [67]. These Ru complexes, in relation to Pt complexes, present better selectivity and lower toxicity. Furthermore, Ru complexes exhibit anti-metastatic behavior, which suggests that Ru compounds inhibit tumor growth and the spread of cancer in advanced stages of the illness [68].

Through the years, organic ligands were researched to improve the inhibition of cancer and the selectivity of tumoral cells. Compound such as NAMI-A (imidazolium trans-[tetrachloro(dimethylsulfoxide)(1H-imidazole)-ruthenate(III)]), KP1019 (indazolium trans-[tetrachlorobis(1H-indazole)ruthenate(III)]), α-[Ru(azpy)$_2$Cl$_2$] and [Ru(H_2cdta)Cl_2] (Figs. 5e, f, h and 6g) is very promising in cancer treatment, and this is because the Ru property of mimic the Fe chemical behavior in metabolism, and organic compounds improve interaction with cytoplasm and posterior penetration inside cell [69–73]. The Ru-complex is primarily applied *in vivo* tests for mammary cancer [55, 73, 74]. Currently, Ru-arene is an essential class of complexes in cancer drug applications. The interest in cymene ligand is because the geometry is a pseudo-octahedron, with a piano-stool shape, which protects the Ru metallic center and stabilizes the redox process of complexes and the vast possibilities of ligands in the Ru atom. The great potential for modulation and customization of ligands in Ru-cymene complexes creates numerous possibilities for properties and applications. The reason for high stabilization in Ru-arene is the interaction between Ru 3d orbitals and 2p delocalized orbitals, characterizing an n-type bond [75–77]. Phosphine ligands are utilized in Ru-complexes for the enormous possibilities of modification in structure. It is reported that the PPh_3 ligand and analogues with modification in Ph bonds present general σ-donor behavior in Ru-complexes. This behavior stabilizes the Ru^{2+} and redox process in $P \rightarrow Ru \rightarrow cy$ charge transfer. Therefore, the cymene ligand, PPh_3, is inert in physiological ambient conditions and presents low toxicity to healthy cells and in metabolism. The PPh_3 ligand, in which it supports many modifications in structure, ensures chemical behavior control and modulation from ligand modification in the Ph structure [78–82].

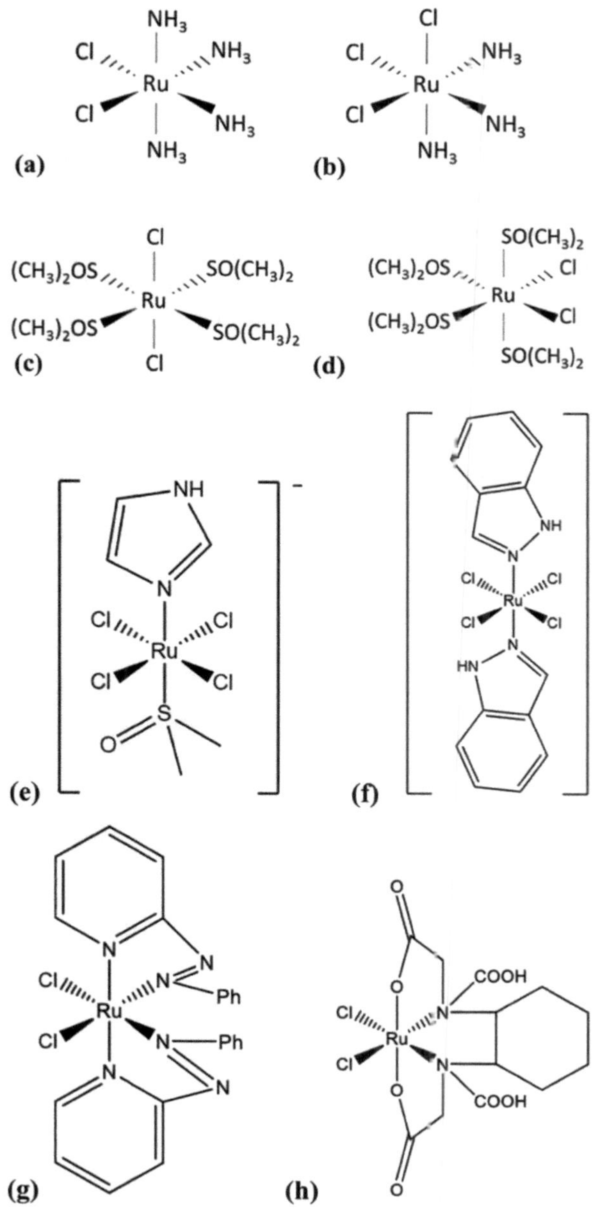

Fig. 5 Molecular structure of Ru-complexes applied on research to application in cancer treatment (**a**) [cis-$RuCl_2(NH_3)_4$]Cl; (**b**) fac-$RuCl_3(NH_3)_3$; (**c**) trans-$RuCl_2$(dimetil-sulfoxido); (**d**) cis-$RuCl_2$(dimethyl sulfoxide); (**e**) NAMI-A (imidazolium trans-[tetrachloro(dimethylsulfoxide)(1H-imidazole)-ruthenate(III)]), (**f**) KP1019 (indazolium trans-[tetrachlorobis(1H-indazole) ruthenate(III)]), (**g**) α-[$Ru(azpy)_2Cl_2$] and (**h**) [$Ru(H_2cdta)Cl_2$]

3 Theoretical DFT Calculation of Ru(II)-η6-p-Cymene-Trifluorbenzylphosfone Complex

In the present theoretical work, the DFT/B3LYP calculation level simulated the Ru(II)-η6-p-cymene-trifluorbenzylphosfone (Ru-cyTF) complex (Fig. 6), where there was a combination of the Ru-cymene complex with the Phosphine ligand [83–86]. The objective of the redox process from frontier orbitals and the Koopman theorem [87–89] evaluated the electronic feature with DNA sites. As widely reported, metallic complexes interact with the double helix of DNA through nitrogenous base pairs that are well-defined in two options: Guanine-Cytosine (G-C) and Adenine-Thymine (A-T) pairs. To verify the interaction affinity based on electronegativity and frontier orbitals, the nitrogenous base pair is simulated, eliminating the need to simulate the entire DNA structure, which results in a lower computational cost. This approximation is supported by other theoretical works that suggest interactions in specific DNA sites, such as the Guanine-Cytosine site [90]. Although the standard approach is the direct interaction of metallic complexes with nitrogenous bases, this can be improved to enhance analysis and reduce computational cost.

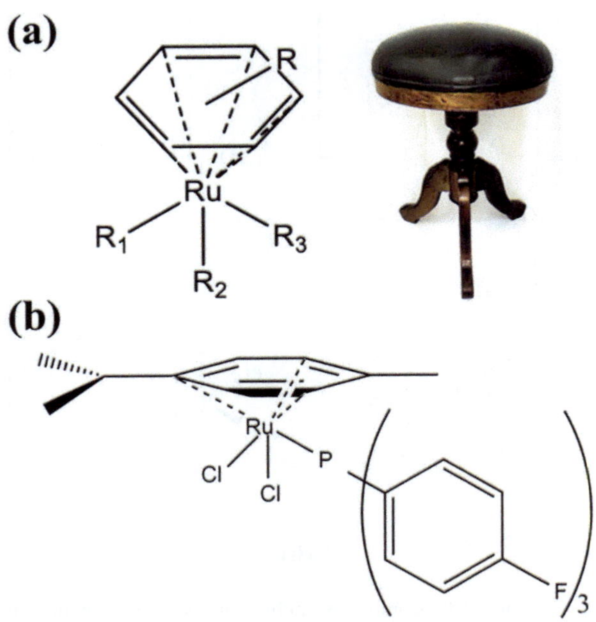

Fig. 6 Molecular structure of (**a**) Ru-(p-cymene) with "piano stool" shape, with Ru atom in pseudo-octahedron geometry, and (**b**) Ru(II)-η6-p-cymene-trifluorbenzylphosfone complex

4 Computational Methods

The theoretical investigation of Ru(II)-η6-p-cymen-trifluorbenzylphosfone (Ru-cyTF) complex utilized a DFT/UB3LYP unrestricted hybrid functional to calculate the electronic exchange-correlation description. The C, H, P, and F atoms are described by 6–311 + G(d) Gaussian functions, and the transition metal Ru by SPK-DZCD Double-Zeta Gaussian basis-set [90]. Calculations are supported in Gaussian09 software package [91, 92]. The structural analysis of the Ru-cyTF complex (Fig. 5) is the measure of $Ru - C\ell_1$, $Ru - C\ell_2$ and $Ru - P$ bond length and $Ru - cim$, that represent the distance of metal to center of aromatic ring in cymene ligand group. The frequency calculations evaluated the minimal energy of the molecules obtained in simulations. Charge -1 simulated the redox behavior of the Ru-cyTF complex (Ru-cyTF^{red}), while the oxidized process was through the charge +1 (Ru-cyTF^{oxi}). The Polarizable Continuum Method (PCM) simulated the solvation effect with water [93]. Frontier Orbitals describes the electronic profile featuring the High Occupied Molecular Orbital (HOMO) and Lowest Unoccupied Molecular Orbital (LUMO) orbitals. From frontier orbitals analyzed the atomic composition on the HOMO and LUMO orbitals among in Ru metallic center, $C\ell$, cymene, and phosphine ligands. The Multiwfm software [95] supported such orbital analysis. Other fundamental properties of the RucyTF complex were Electron Affinity (Eq. 1), Ionization Energy (Eq. 2), Electronegativity (χ) (Eq. 3), and Chemical Hardness (η) (Eq. 4) by Koopman theorem [96, 97]. This characterization is essential to elucidate the electronic profile and stability of the RucyTF-complex regarding oxidation and reduction processes.

$$EA = E_{LUMO} \tag{1}$$

$$IE = -E_{HOMO} \tag{2}$$

$$\chi = \frac{(IE + EA)}{2} \tag{3}$$

$$\eta = \frac{IE - AE}{2} \tag{4}$$

The interaction with DNA structure site occurred on two nitrogenous base pair, the Guanine-Cytosine (G-C) and the Adenine-Thymine (A-T) (Fig. 7). These systems are characterized and electronic behavior compared with RucyTF complex to propose a possible interaction with specify site in DNA structure.

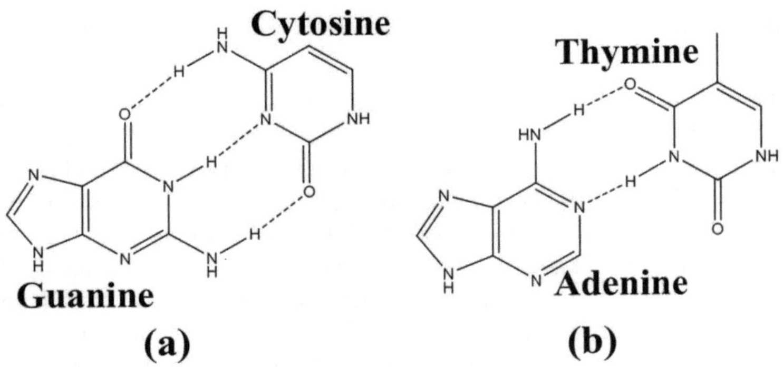

Fig. 7 Nitrogenous base pairs of DNA structure sites to interact with RucyTF complex. (**a**) Guanine-Cytosine, and (**b**) Thymine-Adenine

5 Results and Discussion

The Ru-cyTF molecular geometries in vacuum and under aqueous solution presented similarly with low modulation of the bond lengths (Table 1). The electrostatic field simulation of the aqueous media had weak forces on the Ru-cyTF geometry. For the Ru-cyTF^{oxi} molecule, the positive charge carried out through oxidation process decreased the Ru–Cl lengths and increased the Ru–cy length moving away the cymene ligand. The solvent behavior keeps the bond lengths from vacuum environment. The Ru-cyTF^{red} geometry shows distortions in bond lengths. The most perceptive was Ru–Cl_1 bond length with 4.727 Å indicating the Cl atom as a leaving group. The solvent effect increased the Ru–Cl_1 length for 8.945 Å intensifying the abandon of the chlorine group. There was a charge transference of the negative charge on the Ru-cyTF^{red} to the chlorine group causing a change on the molecular geometry of the Ruthenium complex. The influence of the aqueous solvent on Ru-cyTF^{red} was fundamental to acentuate essential electronic behaviors as leaving group character of the Cl atom predicts in reported organic mechanisms [98]. In general, the solvation of molecule stabilizes the charge transference modifying the Ru-cyTF^{red} to a Ru atom pseudo-octahedron configuration of Ru atom. Reduced molecule was more alterated by solvent effect than the oxided molecule through Ru–Cl bond length.

In analysis of the Charge Density Maps (CDM) (Fig. 8), the Ru-cyTF^{red} in vacuum presented a charge transfer for the chloro atom locating the negative charge. The charge transfer is completed with the solvent effect (Fig. 8b) from the total separation of negative charge with the Cl^- ion . There is a new Ru-cyTF^{red} structure and a neutral profile besides new stoichiometry.

Now, for the Ru-cyTF complex, the charge density is homogenously distributed on the molecular geometry with negative charge on P and Cl atoms as vaccum as solvated systems (Fig. 8c, d). The positive charges are distributed on cyneme and aromatic ring indicating the influence of the electronegativity on the π electrons.

Table 1 Bond lengths analysis for the Ru-cyTF^{oxi}, Ru-cyTF, and Ru-cyTF^{red} complexes in vacuum and solvated aqueous environments

Environment	Bond lengths	Ru-cyTF^{oxi}	Ru-cyTF	Ru-cyTF^{red}
Vacuum	$Ru-C\ell_1$ $\left(\overset{\circ}{A}\right)$	2.332	2.446	4.727
	$Ru-C\ell_2$ $\left(\overset{\circ}{A}\right)$	2.334	2.446	2.443
	$Ru-P$ $\left(\overset{\circ}{A}\right)$	2.407	2.449	2.372
	$Ru-cy$ $\left(\overset{\circ}{A}\right)$	1.971	1.733	1.796
Aqueous	$Ru-C\ell_1$ $\left(\overset{\circ}{A}\right)$	2.332	2.465	8.945
	$Ru-C\ell_2$ $\left(\overset{\circ}{A}\right)$	2.334	2.465	2.492
	$Ru-P$ $\left(\overset{\circ}{A}\right)$	2.407	2.426	2.392
	$Ru-cy$ $\left(\overset{\circ}{A}\right)$	1.971	1.799	1.823

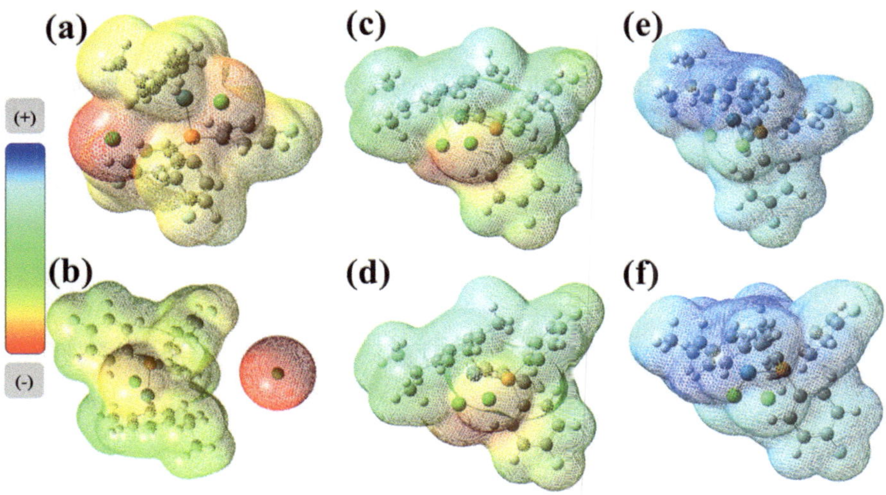

Fig. 8 Charge density maps of (**a**) Ru-cyTF^{red} in vacuum and (**b**) solvated; (**c**) Ru-cyTF in vacuum and (**d**) solvated; (**e**) Ru-cyTF^{oxi} in vacuum and (**f**) solvated. The Grey, white, green, cinan, light blue and orange colors represent C, H, Cl, Ru, F and P atoms. The charge scale represented in left corner represent in blue positive charge, green colors to neutral charge and red to negative charge

Furthermore, both vacuum and aqueous environmets showed low molecular geometry distortion on the Ru-cyTF^{oxi} molecule. The effect of the positive charge is spread on all molecular geometry without a specific location on atom or chemical group.

To follow in more details of the electronic distribution on the Ru-cyTF complexes and the charge transfer mechanism, the contribution of the Ru, Cl, phosphine, and cymene chemical groups on the HOMO and LUMO frontier orbitals are in Table 2. The water solvent in Ru-cyTF had a low effect on orbitals, while the Ru and cymene groups showed more contribution on the HOMO orbital. In LUMO orbital, the Ru and phosphine groups make more contribution. Then, the charge transfer from electrons excitation occurs from cymene to phosphine. In Ru-cyTF neutral, the most contribution in the frontier orbitals are defined by Ru and Cl atom contribution in HOMO orbital. However, in LUMO orbital, the more contribution is between Ru and cymene. The charge transfer in neutral profile from frontier orbitals is from Cl atom to cymene intermediated by Ru. In another way, the oxidation process induced anomalous behavior.

Only the phosphine group contributed significatively on the HOMO orbital. In LUMO orbital, the electronic density is shared between Ru and Cl atoms. This means that in the redox process, Ru atom has important rules to charge transfer in eject or receive electron in extra and intramolecular process. In reduced and neutral molecules, the Ru atom is an intermediary among a ligand through the frontier orbitals. Meanwhile, the Ru atom acts as receptor of the electrons from phosphine in oxidated complex. Such chemical characteristic show that charge transfer mechanism is combined between Ru atom and phosphine or cymene ligand.

In Fig. 9 there are the HOMO and LUMO orbitals under water solvent. All Ru-cyTF complexes present $n \rightarrow \pi$ charge transfer in structure, where n states refer to orbital localized between Ru and cymene ligand. This electronic profile is result of cymene coordination with Ru atom and the redox process of $Ru^{2+} \rightarrow Ru^{3+}$ or

Table 2 Contributions of the Ru metallic center, Cymene, Chloride, and Phosphine groups in the HOMO and LUMO frontier orbitals for the Ru-cyTF^{red}, Ru-cyTF, and Ru-cyTF^{oxi} molecules in water solvent effect

Molecule	Group	HOMO (%)	LUMO (%)
Ru-cyTF^{red}	Ru	51.8	60.3
	Cymene	29.2	15.9
	Cl	16.0	3.7
	Phosphine	3.0	20.0
Ru-cyTF	Ru	42.6	52.9
	Cymene	16.6	24.9
	Cl	39.1	16.5
	Phosphine	4.7	5.7
Ru-cyTF^{oxi}	Ru	2.4	59.2
	Cymene	1.0	13.3
	Cl	3.9	24.6
	Phosphine	92.7	2.9

$Ru^{2+} \rightarrow Ru^{1+}$ depends on Ru-cymene interaction and charge transfer. In reduction process, the charge transfer mechanism is of $cy \rightarrow Ru \rightarrow C\ell$, while in oxidation process is $cy - Ru - Phosphine$. All the mechanisms of charge transfer in Ru-cyTF complexes stabilize the redox process in molecules and preserves the structure. Reduction of Ru-cyTF has labilization of $C\ell$ atom of structure, this process is considerably an irreversible stable state. Beforehand, oxidation presents a reversible stable state.

From frontier orbitals and Koopman theorem, it was analyzed Ionization Energy (IE), Electron Affinity (EA), Electronegativity (χ) and Hardness (η) (Table 3). The IE and AE values presented significant modulations with solvent effct on the Ru-cymTF. By Koopman theorem, this result suggests that molecular orbitals are sensitive to interaction to solvent as reported in theoretical works in literature [99]. The shape of orbital is preserved in electronic structure, but the energy is affected. The IE and AE increased the energy values with the solvent effect causing a resistance to charge transfer. These values indicate an increase in electronegativity and chemical hardness. In relation to η, Ru-cyTF presents chemical profile of soft-polarized. The χ profile presents an increase of 0.652 eV, Ru-cyTF complex presents an increase of chemical behavior to attract electrons and reduce the molecule.

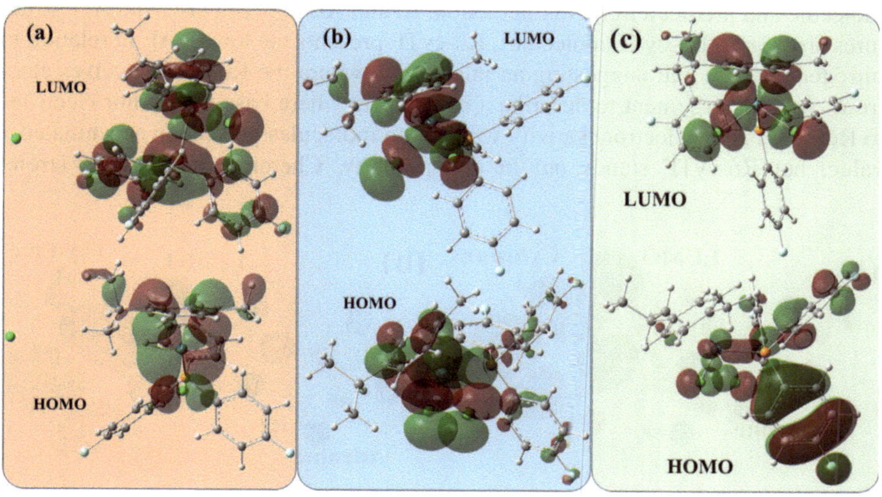

Fig. 9 Frontier molecular orbital for (**a**) Ru-cyTF^{red}, (**b**) Ru-cyTF, (**c**) Ru-cyTF^{oxi} structures with water solvent effect

Table 3 Energetic values of electronic properties such Ionization Energy, Electron Affinity, Electronegativity (χ) and chemical hardness (η) of Ru-cyTF complex in vacuum and solvated in water

	Vacuum	Solvation
Ionization energy (eV)	5.548	6.283
Electron affinity (eV)	−1.791	−1.760
χ (eV)	3.370	4.022
η (eV)	1.879	2.261

In study of interaction with nitrogenous base pair, the interest was the proposition of an interaction with DNA structure through experimental and theoretical data in literature [100–105]. The frontier orbitals of G-C (Fig. 10a) and A-T systems (Fig. 10b) show that the nitrogenous base pair are on the HOMO and LUMO orbitals. In G-C, HOMO orbital is localized in Guanine molecule while LUMO in Cytosine. Similarly, in the A-T system, the HOMO is localized in Adenine and the LUMO in Thymine. In a nutshell, HOMO orbitals are localized in purines molecules of DNA. In reactive terms, the Purines has reductive behavior while Pyrimidines oxidative behavior (Table 3).

The configuration of frontier orbitals is very similar to theoretical results of literature, as reported in Cao and Hou work [106], the frontier orbitals are localized in specifical regions of G-C and A-T sites, which HOMO orbitals majority localizes in purines molecules while LUMO localizes majority in pyrimidines molecules. Is know that in DNA structure that metallic complexes of interest to anti-cancer properties and application in pharmacology interact specifically in Guanine or Adenine molecules. Furthermore, between two base pairs, the most reactive site with metallic complexes is the Guanine-Cytosine in DNA [107–110].

With electronic profile of reactive parameters of Table 4, it is possible to characterize the electronegativity of a specifical base pair and compare it with Ru-cyTF molecule and predict a trend to interaction. IE and AE between G-C and A-T structures are near to Ru-cyTF molecules. Ru-cyTF presents the lowest AE in relation to nitrogenous base pair suggesting an interaction, where the Ru-cyTF receives electrons. Another argument to describe a mechanism where base pair donor electrons to Ru-cyTF, is the electronegativity value, both molecules present an accentuated χ value, but Ru-cyTF stands out in this property. Chemical hardness is strong

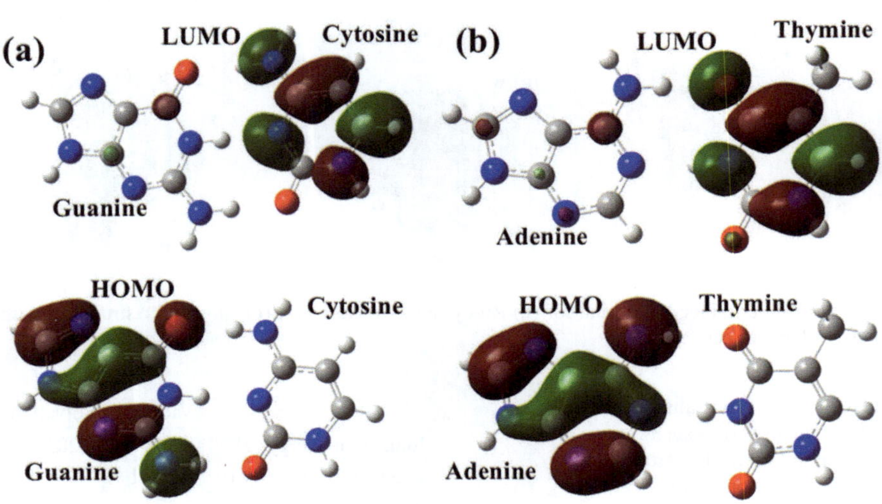

Fig. 10 Frontier molecular orbital for nitrogenous base pair (**a**) Guanine-Cytosine and (**b**) Adenine-Thymine structures with solvation in water

Table 4 Energetic values of electronic properties such ionization energy, electron affinity, electronegativity (χ) and chemical hardness (η) of Ru-cyTF, nitrogenous base pair Guanine-Cytosine (G-C) and Adenine-Thymine (A-T), both solvated in water

	G–C	A–T	Ru-cyTF
Ionization energy (eV)	5.929	6.364	6.283
Electron affinity (eV)	−1.393	−1.426	−1.760
χ (eV)	3.661	3.895	4.022
η (eV)	2.268	2.468	2.261

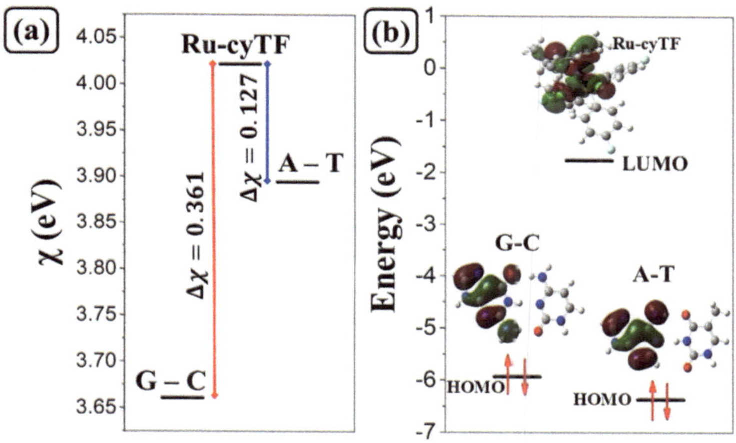

Fig. 11 Energetic diagram comparative to analyze (**a**) electronegativity values and (**b**) frontier orbitals HOMO for Guanine-Cytosine (G-C) and Adenine-Thymine (A-T) systems and LUMO to Ru-cyTF

evidence to determine these interactions as effective interaction from an electronic point of view. This because η has close values with G-C and A-T systems. In another words, the chemical hardness evidences the electronic polarization in both molecules stabilizing the interaction. The energetic approximation between A-T and G-C frontier orbitals with Ru-cyTF frontier orbitals indicates the coupling. To specify such interaction, carried out the analyses of the χ and HOMO location between nitrogenous base pair and Ru-cyTF LUMO orbital predict the site of interaction (Fig. 11). In relation to χ value (Fig. 11a), the highest difference between structures suggests a most effective electronic interaction and reaction to charge transfer.

With this behavior pointed out, it is predicted that G-C system has the most effective interaction with Ru-cyTF complex. The η value indicates that the G-C system is nearest to Ru-cyTF that A-T. Electronic polarization has a low difference of just 0.007 eV, while A-T system presents energetic difference of 0.207 eV. While in orbitals energy (Fig. 11b), G-C and A-T systems are close HOMO energy values. However, LUMO energy for Ru-cyTF and the G-C system are nearest energy. These analysis of chemical profile and electronic properties suggest that the interaction between nitrogenous base pairs and Ru-cyTF complex has higher efficiency with

G-C site, and this is coherent with theoretical and experimental data in literature, which complexes are reduced by DNA structure and posterior bond formation Ru-Guanine structure.

6 Conclusion

The Ru-cyTF molecule is stable to redox process and preserves the pseudo-octahedron configuration to Ru atom. In reduction and oxidation process, Ru atom acts as main intermediate to charge transfer in complex. In the oxidation process, the charge transfer occurs from Cl atoms to Ru atoms in the process $Ru^{2+} \rightarrow Ru^{3+}$ and transfers to phosphine ligand. This mechanism is reversible, because the molecular structure is preserved without significant distortions. Meanwhile, the reduction of Ru-cyTF is irreversible because the process presents Cl as leaving group. The charge transfer mechanism in structure is Cl ligand to Ru and cymene ligand, with reduction of metallic center $Ru^{2+} \rightarrow Ru^{1+}$. The redox stability of Ru-cyTF is coherent with potential catalyst behavior of Ru-complexes. The solvent does not have effect significantly the frontier orbital distribution compared with vacuum system. Although, the orbital energy and consequently electronic properties present expressive variations but preserve the electronic behavior of molecules. The G-C and A-T systems have planar structure that is expected in DNA structure. HOMO orbital of the nitrogenous base pair is localized in the purine. In the interaction study, the G-C is the most provable reductor site from the indirect analysis of reduction between Ru-cyTF and DNA structure through Guanine group.

References

1. Perera, F.P.: Environment and cancer: who are susceptible? Science. **278**, 1068–1073 (1997). https://doi.org/10.1126/science.278.5340.1068
2. Venitt, S.: Mechanisms of carcinogenesis and individual susceptibility to cancer. Clin. Chem. **40**, 1421–1425 (1994). https://doi.org/10.1093/clinchem/40.7.1421
3. Ma, W., Wu, H., Chen, Y., Xu, H., Jiang, J., Du, B., et al.: New techniques to identify the tissue of origin for cancer of unknown primary in the era of precision medicine: progress and challenges. Brief. Bioinform. **25**, bbae028 (2024). https://doi.org/10.1093/bib/bbae028
4. Andreani, T., Cheng, R., Elbadri, K., Ferro, C., Menezes, T., dos Santos, M.R., et al.: Natural compounds-based nanomedicines for cancer treatment: future directions and challenges. Drug Deliv. Transl. Res. **14**, 2845–2916 (2024). https://doi.org/10.1007/s13346-024-01649-z
5. Rai, H.M.: Cancer detection and segmentation using machine learning and deep learning techniques: a review. Multimed. Tools Appl. **83**, 27001–27035 (2023). https://doi.org/10.1007/s11042-023-16520-5
6. Matthews, H.K., Bertoli, C., de Bruin, R.A.M.: Cell cycle control in cancer. Nat. Rev. Mol. Cell Biol. **23**, 74–88 (2022). https://doi.org/10.1038/s41580-021-00404-3
7. Rumgay, H., Shield, K., Charvat, H., Ferrari, P., Sornpaisarn, B., Obot, I., et al.: Global burden of cancer in 2020 attributable to alcohol consumption: a population-based study. Lancet Oncol. **22**, 1071–1080 (2021). https://doi.org/10.1016/S1470-2045(21)00279-5

8. Brown, K.A.: Metabolic pathways in obesity-related breast cancer. Nat. Rev. Endocrinol. **17**, 350–363 (2021). https://doi.org/10.1038/s41574-021-00487-0
9. Kheifets, L.I., Matkin, C.C.: Industrialization, electromagnetic fields, and breast cancer risk. Environ. Health Perspect. **107**, 145–154 (1999). https://doi.org/10.1289/ehp.99107s1145
10. Gabet, S., Lemarchand, C., Guénel, P., Slama, R.: Breast cancer risk in association with atmospheric pollution exposure: a meta-analysis of effect estimates followed by a health impact assessment. Environ. Health Perspect. **129**, 057012-1–057012-18 (2021). https://doi.org/10.1289/EHP8419
11. Yu, F., Thayer, M., Qasemi, E., Zhu, K., Assadi, A.: Deep learning features in atmospheric chemistry: prediction of cancer morbidity due to air pollution. In: 2017 International Conference on Computational Science and Computational Intelligence (CSCI), pp. 1764–1766. IEEE, New York (2017). https://doi.org/10.1109/CSCI.2017.307
12. Bray, F., Laversanne, M., Sung, H., Ferlay, J., Siegel, R.L., Soerjomataram, I., et al.: Global cancer statistics 2022: GLOBOCAN estimates of incidence and mortality worldwide for 36 cancers in 185 countries. CA Cancer J. Clin. **74**, 229–263 (2024). https://doi.org/10.3322/caac.21834
13. Siegel, R.L., Giaquinto, A.N., Jemal, A.: Cancer statistics, 2024. CA Cancer J. Clin. **74**, 12–49 (2024). https://doi.org/10.3322/caac.21820
14. Chen, H., Gao, Y., Wang, A., Zhou, X., Zheng, Y., Zhou, J.: Evolution in medicinal chemistry of ursolic acid derivatives as anticancer agents. Eur. J. Med. Chem. **92**, 648–655 (2015). https://doi.org/10.1016/j.ejmech.2015.01.031
15. Basourakos, S.P., Li, L., Aparicio, A.M., Corn, P.G., Kim, J., Thompson, T.C.: Combination platinum-based and DNA damage response-targeting cancer therapy: evolution and future directions. Curr. Med. Chem. **24**, 1586–1606 (2017). https://doi.org/10.2174/0929867323666161214114948
16. McQuade, R.M., Stojanovska, V., Bornstein, J.C., Nurgali, K.: Colorectal cancer chemotherapy: the evolution of treatment and new approaches. Curr. Med. Chem. **24**, 1537–1557 (2017). https://doi.org/10.2174/0929867324666170111152436
17. Fidler, M.M., Bray, F., Soerjomataram, I.: The global cancer burden and human development: a review. Scand. J. Public Health. **46**, 27–36 (2018). https://doi.org/10.1177/1403494817715400
18. Coussens, L.M., Werb, Z.: Inflammation and cancer. Nature. **420**, 860–867 (2022). https://doi.org/10.1038/nature01322
19. Watson, J.D., Crick, F.H.C.: The structure of DNA. Cold Spring Harb. Symp. Quant. Biol. **18**, 123–131 (1953). https://doi.org/10.1101/SQB.1953.018.01.020
20. Li, J., Chiu, T.-P., Rohs, R.: Predicting DNA structure using a deep learning method. Nat. Commun. **15**, 1243 (2024). https://doi.org/10.1038/s41467-024-45191-5
21. Koehrer, S., Burger, J.A.: Chronic lymphocytic leukemia: disease biology. Acta Haematol. **147**, 8–21 (2024). https://doi.org/10.1159/000533610
22. Feldman, R.P., Goodrich, J.T.: The Edwin Smith surgical papyrus. Childs Nerv. Syst. **15**, 281–284 (1999). https://doi.org/10.1007/s003810050395
23. de Moulin, D.: A short history of breast cancer. Springer Netherlands, Dordrecht (1989). https://doi.org/10.1007/978-94-009-1059-1
24. Karpozilos, A., Pavlidis, N.: The treatment of cancer in Greek antiquity. Eur. J. Cancer. **40**, 2033–2040 (2004). https://doi.org/10.1016/j.ejca.2004.04.036
25. Ekmektzoglou, K.A., Xanthos, T., German, V., Zografos, G.C.: Breast cancer: from the earliest times through to the end of the 20th century. Eur. J. Obstet. Gynecol. Reprod. Biol. **145**, 3–8 (2009). https://doi.org/10.1016/j.ejogrb.2009.03.017
26. Bland, K.I., Copeland, E.M., Klimberg, V.S.: History of the therapy of breast cancer. In: The breast, vol. 3, pp. 1–19. Elsevier, Amsterdam (2018). https://doi.org/10.1016/B978-0-323-35955-9.00001-5
27. Sheppard, J.: Illustrations from the Wellcome Institute Library. Charles Joseph Singer, DM, DLitt, DSc, FRCP (1876–1960): papers in the Contemporary Medical Archives Centre. Med. Hist. **31**, 466–471 (1987). https://doi.org/10.1017/S002572730004730X

28. Gensini, G.F., Conti, A.A., Lippi, D.: The contributions of Paul Ehrlich to infectious disease. J. Infect. **54**, 221–224 (2007). https://doi.org/10.1016/j.jinf.2004.05.022
29. Chang, A.E.: Oncology. Springer, New York, NY (2006). https://doi.org/10.1007/0-387-31056-8
30. Kaufmann, S.H.E.: Paul Ehrlich: founder of chemotherapy. Nat. Rev. Drug Discov. **7**, 373 (2008). https://doi.org/10.1038/nrd2582
31. Valent, P., Groner, B., Schumacher, U., Superti-Furga, G., Busslinger, M., Kralovics, R., et al.: Paul Ehrlich (1854-1915) and his contributions to the foundation and birth of translational medicine. J. Innate Immun. **8**, 111–120 (2016). https://doi.org/10.1159/000443526
32. Huang, W., Zeng, Y.C.: A candidate for lung cancer treatment: arsenic trioxide. Clin. Transl. Oncol. **21**, 1115–1126 (2019). https://doi.org/10.1007/s12094-019-02054-6
33. Waxman, S., Anderson, K.C.: History of the development of arsenic derivatives in cancer therapy. Oncologist. **6**, 3–10 (2001). https://doi.org/10.1634/theoncologist.6-suppl_2-3
34. Hannuksela-Svahn, A., Pukkala, E., Läärä, E., Poikolainen, K., Karvonen, J.: Psoriasis, its treatment, and cancer in a cohort of Finnish patients. J. Invest. Dermatol. **114**, 587–590 (2000). https://doi.org/10.1046/j.1523-1747.2000.00898.x
35. Hughes, M.F.: Arsenic toxicity and potential mechanisms of action. Toxicol. Lett. **133**, 1–16 (2002). https://doi.org/10.1016/S0378-4274(02)00084-X
36. Ghabili, K., Agutter, P.S., Ghanei, M., Ansarin, K., Shoja, M.M.: Mustard gas toxicity: the acute and chronic pathological effects. J. Appl. Toxicol. **30**, 627–643 (2010). https://doi.org/10.1002/jat.1581
37. Adair, F.E., Bagg, H.J.: Experimental and clinical studies on the treatment of cancer by dichlorethylsulphide (mustard gas). Ann. Surg. **93**, 190–199 (1931). https://doi.org/10.1097/00000658-193101000-00026
38. Wiedemann, H.R.: Sidney Farber. Eur. J. Pediatr. **153**, 223 (1994). https://doi.org/10.1007/BF01954506
39. MacAulay, M.: Book review: Eula Biss, on immunity: an inoculation. Health. **23**, 576–577 (2019). https://doi.org/10.1177/1363459316656309
40. Blattner, R.J., Williamson, A.P., Simonsen, L., Robertson, G.G.: Teratogenesis with cancer chemotherapeutic agents. J. Pediatr. **56**, 285–293 (1960). https://doi.org/10.1016/S0022-3476(60)80127-8
41. Desmoulin, S.K., Hou, Z., Gangjee, A., Matherly, L.H.: The human proton-coupled folate transporter. Cancer Biol. Ther. **13**, 1355–1373 (2012). https://doi.org/10.4161/cbt.22020
42. Farber, S., Diamond, L.K., Mercer, R.D., Sylvester, R.F., Wolff, J.A.: Temporary remissions in acute leukemia in children produced by folic acid antagonist, 4-aminopteroyl-glutamic acid (aminopterin). N. Engl. J. Med. **238**, 787–793 (1948). https://doi.org/10.1056/NEJM194806032382301
43. Farber, S.: Some observations on the effect of folic acid antagonists on acute leukemia and other forms of incurable cancer. Blood. **4**, 160–167 (1949). https://doi.org/10.1182/blood.V4.2.160.160
44. Tang, C., Livingston, M.J., Safirstein, R., Dong, Z.: Cisplatin nephrotoxicity: new insights and therapeutic implications. Nat. Rev. Nephrol. **19**, 53–72 (2023). https://doi.org/10.1038/s41581-022-00631-7
45. Zhang, J., Ye, Z., Tew, K.D., Townsend, D.M.: Cisplatin chemotherapy and renal function. Adv. Cancer Res. **152**, 305–327 (2021). https://doi.org/10.1016/bs.acr.2021.03.008
46. Cohen, S.M., Lippard, S.J.: Cisplatin: from DNA damage to cancer chemotherapy. Prog. Nucleic Acid Res. Mol. Biol. **61**, 93–130 (2021). https://doi.org/10.1016/S0079-6603(01)67026-0
47. Zoń, A., Bednarek, I.: Cisplatin in ovarian cancer treatment—known limitations in therapy force new solutions. Int. J. Mol. Sci. **24**, 7585 (2023). https://doi.org/10.3390/ijms24087585
48. Hellesnes, R., Myklebust, T.Å., Fosså, S.D., Bremnes, R.M., Karlsdottir, Á., Kvammen, Ø., et al.: Testicular cancer in the cisplatin era: causes of death and mortality rates in a population-based cohort. J. Clin. Oncol. **39**, 3561–3573 (2021). https://doi.org/10.1200/JCO.21.00637

49. Romani, A.M.P.: Cisplatin in cancer treatment. Biochem. Pharmacol. **206**, 115323 (2022). https://doi.org/10.1016/j.bcp.2022.115323
50. Stefanache, A., Miftode, A.M., Constantin, M., Bogdan Goroftei, R.E., Olaru, I., Gutu, C., et al.: Noble metal complexes in cancer therapy: unlocking redox potential for next-gen treatments. Inorganics (Basel). **13**, 64 (2025). https://doi.org/10.3390/inorganics13020064
51. Smith, C.B., Days, L.C., Alajroush, D.R., Faye, K., Khodour, Y., Beebe, S.J., et al.: Photodynamic therapy of inorganic complexes for the treatment of cancer. Photochem. Photobiol. **98**, 17–41 (2022). https://doi.org/10.1111/php.13467
52. Kumar Singh, A., Kumar, A., Singh, H., Sonawane, P., Pathak, P., Grishina, M., et al.: Metal complexes in cancer treatment: journey so far. Chem. Biodivers. **20**, e202300061 (2023). https://doi.org/10.1002/cbdv.202300061
53. Gourdon, L., Cariou, K., Gasser, G.: Phototherapeutic anticancer strategies with first-row transition metal complexes: a critical review. Chem. Soc. Rev. **51**, 1167–1195 (2022). https://doi.org/10.1039/D1CS00609F
54. Sun, Q., Li, Y., Shi, H., Wang, Y., Zhang, J., Zhang, Q.: Ruthenium complexes as promising candidates against lung cancer. Molecules. **26**, 4389 (2021). https://doi.org/10.3390/molecules26154389
55. Liang, J.-X., Zhong, H.-J., Yang, G., Vellaisamy, K., Ma, D.-L., Leung, C.-H.: Recent development of transition metal complexes with in vivo antitumor activity. J. Inorg. Biochem. **177**, 276–286 (2017). https://doi.org/10.1016/j.jinorgbio.2017.06.002
56. Mahmud, K.M., Niloy, M.S., Shakil, M.S., Islam, M.A.: Ruthenium complexes: an alternative to platinum drugs in colorectal cancer treatment. Pharmaceutics. **13**, 1295 (2021). https://doi.org/10.3390/pharmaceutics13081295
57. Lu, Y., Zhu, D., Le, Q., Wang, Y., Wang, W.: Ruthenium-based antitumor drugs and delivery systems from monotherapy to combination therapy. Nanoscale. **14**, 16339–16375 (2022). https://doi.org/10.1039/D2NR02994D
58. Sumithaa, C., Ganeshpandian, M.: Half-sandwich ruthenium arene complexes bearing clinically approved drugs as ligands: the importance of metal–drug synergism in metallodrug design. Mol. Pharm. **20**, 1453–1479 (2023). https://doi.org/10.1021/acs.molpharmaceut.2c01027
59. Ishizuka, T., Hosokawa, A., Kawanishi, T., Kotani, H., Zhi, Y., Kojima, T.: Self-photosensitizing dinuclear ruthenium catalyst for CO2 reduction to CO. J. Am. Chem. Soc. **145**, 23196–23204 (2023). https://doi.org/10.1021/jacs.3c07685
60. Li, Y., Qin, T., Wei, Y., Xiong, J., Zhang, P., Lai, K., et al.: A single site ruthenium catalyst for robust soot oxidation without platinum or palladium. Nat. Commun. **14**, 7149 (2023). https://doi.org/10.1038/s41467-023-42935-7
61. Ye, C.-X., Meggers, E.: Chiral-at-ruthenium catalysts for nitrene-mediated asymmetric C–H functionalizations. Acc. Chem. Res. **56**, 1128–1141 (2023). https://doi.org/10.1021/acs.accounts.3c00081
62. Shin, S., Kang, D.C., Kim, K., Jeong, Y., Kim, J., Lee, S., et al.: Emulating the short-term plasticity of a biological synapse with a ruthenium complex-based organic mixed ionic–electronic conductor. Mater. Adv. **3**, 2827–2837 (2022). https://doi.org/10.1039/D1MA01078F
63. Mauri, L., Colombo, A., Dragonetti, C., Roberto, D., Fagnani, F.: Recent investigations on thiocyanate-free ruthenium(II) 2,2′-bipyridyl complexes for dye-sensitized solar cells. Molecules. **26**, 7638 (2021). https://doi.org/10.3390/molecules26247638
64. Avisar, S., Shvets, A., Shner, Y., Popov, I., Bino, A.: Nano-porous ruthenium-palladium and ruthenium-platinum alloys and their application as hydrogenation catalysts. J. Alloys Compd. **936**, 168326 (2023). https://doi.org/10.1016/j.jallcom.2022.168326
65. Sadique, S., Baqer, A.A., Salman, A.W., Iqbal, M.A., Kadim, M.M., Jamil, F., et al.: Ruthenium complexes for breast cancer therapy. Rev. Inorg. Chem. **44**, 191–208 (2024). https://doi.org/10.1515/revic-2023-0010

66. Shen, J., Rees, T.W., Ji, L., Chao, H.: Recent advances in ruthenium(II) and iridium(III) complexes containing nanosystems for cancer treatment and bioimaging. Coord. Chem. Rev. **443**, 214016 (2021). https://doi.org/10.1016/j.ccr.2021.214016
67. Pete, S., Roy, N., Kar, B., Paira, P.: Construction of homo and heteronuclear Ru(II), Ir(III) and Re(I) complexes for target specific cancer therapy. Coord. Chem. Rev. **460**, 214462 (2022). https://doi.org/10.1016/j.ccr.2022.214462
68. Kanaoujiya, R., Meenakshi, Srivastava, S., Singh, R., Mustafa, G.: Recent advances and application of ruthenium complexes in tumor malignancy. Mater. Today Proc. **72**, 2822–2827 (2023). https://doi.org/10.1016/j.matpr.2022.07.098
69. Vilaplana, R.A., González-Vílchez, F., Gutierrez-Puebla, E., Ruiz-Valero, C.: The first isolated antineoplastic Ru(IV) complex: synthesis and structure of [Cl2(1,2-cyclohexanediaminotetraacetate)Ru]·2H2O. Inorg. Chim. Acta. **224**, 15–18 (1994). https://doi.org/10.1016/0020-1693(94)04159-8
70. Hotze, A.C.G., Velders, A.H., Ugozzoli, F., Biagini-Cingi, M., Manotti-Lanfredi, A.M., Haasnoot, J.G., et al.: Synthesis, characterization, and crystal structure of α-[Ru(azpy)2(NO3)2] (azpy = 2-(Phenylazo)pyridine) and the products of its reactions with guanine derivatives. Inorg. Chem. **39**, 3838–3844 (2000). https://doi.org/10.1021/ic000066v
71. Hotze, A.C.G., Kooijman, H., Spek, A.L., Haasnoot, J.G., Reedijk, J.: Synthesis and characterization of ruthenium(II) complexes with the new ligand 2-phenylazopyridine-5-sulfonic acid (Hsazpy): in search for new anticancer agents. New J. Chem. **28**, 565 (2004). https://doi.org/10.1039/b313746e
72. Shah, P.K., Jena, N.R., Shukla, P.K.: Reactions of Ru(III)-drugs KP1019 and KP418 with guanine, 2′-deoxyguanosine and guanosine: a DFT study. J. Mol. Model. **28**, 291 (2022). https://doi.org/10.1007/s00894-022-05304-7
73. Silvestri, S., Cirilli, I., Marcheggiani, F., Dludla, P., Lupidi, G., Pettinari, R., et al.: Evaluation of anticancer role of a novel ruthenium(II)-based compound compared with NAMI-A and cisplatin in impairing mitochondrial functionality and promoting oxidative stress in triple negative breast cancer models. Mitochondrion. **56**, 25–34 (2021). https://doi.org/10.1016/j.mito.2020.11.004
74. Frezza, M., Dou, Q.P., Xiao, Y., Samouei, H., Rashidi, M., Samari, F., et al.: In vitro and in vivo antitumor activities and DNA binding mode of five coordinated cyclometalated organoplatinum(II) complexes containing biphosphine ligands. J. Med. Chem. **54**, 6166–6176 (2011). https://doi.org/10.1021/jm2006832
75. Nowak-Sliwinska, P., van Beijnum, J.R., Casini, A., Nazarov, A.A., Wagnières, G., van den Bergh, H., et al.: Organometallic ruthenium(II) arene compounds with antiangiogenic activity. J. Med. Chem. **54**, 3895–3902 (2011). https://doi.org/10.1021/jm2002074
76. Haribabu, J., Srividya, S., Umapathi, R., Gayathri, D., Venkatesu, P., Bhuvanesh, N., et al.: Enhanced anticancer activity of half-sandwich Ru(II)-p-cymene complex bearing heterocyclic hydrazone ligand. Inorg. Chem. Commun. **119**, 108054 (2020). https://doi.org/10.1016/j.inoche.2020.108054
77. Sumithaa, C., Manjunathan, T., Mazuryk, O., Peters, S., Pillai, R.S., Brindell, M., et al.: Nanoencapsulation of Ru(p-cymene) complex bearing ginger-based natural product into liposomal nanoformulation to improve its cellular uptake and antiproliferative activity. ACS Appl. Bio Mater. **5**, 3241–3256 (2022). https://doi.org/10.1021/acsabm.2c00231
78. Haribabu, J., Balakrishnan, N., Swaminathan, S., Peter, J., Gayathri, D., Echeverria, C., et al.: Synthesis, cytotoxicity and docking studies (with SARS-CoV-2) of water-soluble binuclear Ru-p-cymene complex holding indole thiosemicarbazone ligand. Inorg. Chem. Commun. **134**, 109029 (2021). https://doi.org/10.1016/j.inoche.2021.109029
79. Wrobel, E.C., Guimarães, I.D.L., Wohnrath, K., Oliveira, O.N.: Effects induced by η6-p-cymene ruthenium(II) complexes on Langmuir monolayers mimicking cancer and healthy cell membranes do not correlate with their toxicity. Biochim. Biophys. Acta Biomembr. **1866**, 184332 (2024). https://doi.org/10.1016/j.bbamem.2024.184332

80. Dimbarre Lao Guimarães, I., Marszaukowski, F., Buhrer Rutka, P., Felipe Borge, L., Augusto Pontes Ribeiro, R., Ricardo de Lazaro, S., et al.: Synthesis, characterization and anticancer activities of cationic η6-p-cymene ruthenium(II) complexes containing phosphine and nitrogenous ligands. Polyhedron. **224**, 115980 (2022). https://doi.org/10.1016/j.poly.2022.115980
81. Guimarães, I.D.L., Marszaukowski, F., Ribeiro, R., de Lazaro, S.R., de Oliveira, K.M., Batista, A.A., et al.: Synthesis and characterization of η6-p-cymene ruthenium(II) complexes containing alkyl- and methoxy-substituted triarylphosphines. J. Organomet. Chem. **931**, 121599 (2021). https://doi.org/10.1016/j.jorganchem.2020.121599
82. do Rosario, V.G.B., Marszaukowski, F., Dimbarre Lao Guimarães, I., Gobbo Maranha, F., Francieli Mika, B., Bonifácio Rosa, G., et al.: Synthesis, structure and biological evaluation as antibacterial agents of Ru(II)-p-cymene-aryldicyclohexylphosphine complexes. Inorg. Chim. Acta. **558**, 121749 (2023). https://doi.org/10.1016/j.ica.2023.121749
83. da Silva, J.P., Fuganti, O., Kramer, M.G., Facchin, G., Aquino, L.E.N., Ellena, J., et al.: Electrochemical, mechanistic, and DFT studies of amine derived diphosphines containing Ru(II)–cymene complexes with potent in vitro cytotoxic activity against HeLa and triple-negative breast cancer cells MDA-MB-231. Dalton Trans. **49**, 16498–16514 (2020). https://doi.org/10.1039/D0DT02500C
84. Lu, T.-F., Li, W., Bai, F.-Q., Jia, R., Chen, J., Zhang, H.-X.: Anionic ancillary ligands in cyclometalated Ru(II) complex sensitizers improve photovoltaic efficiency of dye-sensitized solar cells: insights from theoretical investigations. J. Mater. Chem. A Mater. **5**, 15567–15577 (2017). https://doi.org/10.1039/C7TA03360E
85. Escudero, D., González, L.: RASPT2/RASSCF vs range-separated/hybrid DFT methods: assessing the excited states of a Ru(II)bipyridyl complex. J. Chem. Theory Comput. **8**, 203–213 (2012). https://doi.org/10.1021/ct200640q
86. Sieffert, N., Bühl, M.: Noncovalent interactions in a transition-metal triphenylphosphine complex: a density functional case study. Inorg. Chem. **48**, 4622–4624 (2009). https://doi.org/10.1021/ic900347e
87. Backler, F., Wilson, G.J., Wang, F.: Rational use of ligand to shift the UV–vis spectrum of Ru-complex sensitiser dyes for DSSC applications. Radiat. Phys. Chem. **161**, 66–71 (2019). https://doi.org/10.1016/j.radphyschem.2019.01.012
88. Hemmati, R., Mostafanejad, M., Ortiz, J.V.: Numerical analysis of the complete active-space extended Koopmans's theorem. J. Chem. Phys. **161**, 094101 (2024). https://doi.org/10.1063/5.0226057
89. Manjanath, A., Sahara, R., Ohno, K., Kawazoe, Y.: Non-adiabatic excited-state time-dependent GW molecular dynamics (TD GW) satisfying extended Koopmans' theorem: an accurate description of methane photolysis. J. Chem. Phys. **160**, 184102 (2024). https://doi.org/10.1063/5.0202590
90. Fonseca Guerra, C., Bickelhaupt, F.M., Snijders, J.G., Baerends, E.J.: Hydrogen bonding in DNA base pairs: reconciliation of theory and experiment. J. Am. Chem. Soc. **122**, 4117–4128 (2000). https://doi.org/10.1021/ja993262d
91. Noro, T., Sekiya, M., Koga, T.: Segmented contracted basis sets for atoms H through Xe: Sapporo-(DK)-nZP sets (n = D, T, Q). Theor. Chem. Acc. **131**, 1124 (2012). https://doi.org/10.1007/s00214-012-1124-z
92. Remya, K., Suresh, C.H.: Which density functional is close to CCSD accuracy to describe geometry and interaction energy of small noncovalent dimers? A benchmark study using Gaussian09. J. Comput. Chem. **34**, 1341–1353 (2013). https://doi.org/10.1002/jcc.23263
93. Pluhařová, E., Schroeder, C., Seidel, R., Bradforth, S.E., Winter, B., Faubel, M., et al.: Unexpectedly small effect of the DNA environment on vertical ionization energies of aqueous nucleobases. J. Phys. Chem. Lett. **4**, 3766–3769 (2013). https://doi.org/10.1021/jz402106h
94. Schroeder, C.A., Pluhařová, E., Seidel, R., Schroeder, W.P., Faubel, M., Slavíček, P., et al.: Oxidation half-reaction of aqueous nucleosides and nucleotides via photoelectron spectroscopy augmented by ab initio calculations. J. Am. Chem. Soc. **137**, 201–209 (2015). https://doi.org/10.1021/ja508149e

95. Slavíček, P., Winter, B., Faubel, M., Bradforth, S.E., Jungwirth, P.: Ionization energies of aqueous nucleic acids: photoelectron spectroscopy of pyrimidine nucleosides and ab initio calculations. J. Am. Chem. Soc. **131**, 6460–6467 (2009). https://doi.org/10.1021/ja8091246
96. Lu, T., Chen, F.W.: Calculation of molecular orbital composition. Acta Chim. Sin. **69**, 2393 (2011)
97. Dong, X., Oganov, A.R., Cui, H., Zhou, X.-F., Wang, H.-T.: Electronegativity and chemical hardness of elements under pressure. Proc. Natl. Acad. Sci. **119**, e2117416119 (2022). https://doi.org/10.1073/pnas.2117416119
98. Luo, J., Xue, Z.Q., Liu, W.M., Wu, J.L., Yang, Z.Q.: Koopmans' theorem for large molecular systems within density functional theory. J. Phys. Chem. A. **110**, 12005–12009 (2006). https://doi.org/10.1021/jp063669m
99. Gibson, D.H., Sleadd, B.A., Mashuta, M.S., Richardson, J.F.: Chloride labilization resulting from nucleophilic addition to cis-Ru(tpy)(CO)2Cl+PF6-: synthesis and characterization of new CO2-bridged and formate complexes of ruthenium. Organometallics. **16**, 4421–4427 (1997). https://doi.org/10.1021/om970493p
100. Mora-Diez, N., Montero, L.A., Fabian, J.: Molecular orbital modeling of solvent effects on excited states of organic molecules. J. Mol. Struct. Theochem. **453**, 49–57 (1998). https://doi.org/10.1016/S0166-1280(98)00186-9
101. Burrows, C.J., Rokita, S.E.: Recognition of guanine structure in nucleic acids by nickel complexes. Acc. Chem. Res. **27**, 295–301 (1994). https://doi.org/10.1021/ar00046a002
102. Arjmand, F., Sayeed, F., Muddassir, M.: Synthesis of new chiral heterocyclic Schiff base modulated Cu(II)/Zn(II) complexes: their comparative binding studies with CT-DNA, mononucleotides and cleavage activity. J. Photochem. Photobiol. B. **103**, 166–179 (2011). https://doi.org/10.1016/j.jphotobiol.2011.03.001
103. Liu, H.-K., Sadler, P.J.: Metal complexes as DNA intercalators. Acc. Chem. Res. **44**, 349–359 (2011). https://doi.org/10.1021/ar100140e
104. Grover, N., Welch, T.W., Fairley, T.A., Cory, M., Thorp, H.H.: Covalent binding of aquaruthenium complexes to DNA. Inorg. Chem. **33**, 3544–3548 (1994). https://doi.org/10.1021/ic00094a018
105. Pages, B.J., Ang, D.L., Wright, E.P., Aldrich-Wright, J.R.: Metal complex interactions with DNA. Dalton Trans. **44**, 3505–3526 (2015). https://doi.org/10.1039/C4DT02700K
106. Cao, G.-J., Hou, H.-L.: Dinuclear metal-mediated guanine–uracil base pairs: theoretical studies of GUM22+ (M = Cu, Ag, and Au) ions. J. Clust. Sci. **30**, 439–448 (2019). https://doi.org/10.1007/s10876-019-01503-0
107. Cao, G.-J.: Dinuclear metal-mediated homo base pairs with metallophilic interactions: theoretical studies of G2M2 2+ (M = Cu, Ag, and Au) ions. Sci. Rep. **7**, 14896 (2017). https://doi.org/10.1038/s41598-017-14259-2
108. Robertazzi, A., Platts, J.A.: Binding of transition metal complexes to guanine and guanine–cytosine: hydrogen bonding and covalent effects. JBIC J. Biol. Inorg. Chem. **10**, 854–866 (2005). https://doi.org/10.1007/s00775-005-0034-0
109. Fuentes-Cabrera, M., Sumpter, B.G., Šponer, J.E., Šponer, J., Petit, L., Wells, J.C.: Theoretical study on the structure, stability, and electronic properties of the guanine–Zn–cytosine base pair in M-DNA. J. Phys. Chem. B. **111**, 870–879 (2007). https://doi.org/10.1021/jp066465e
110. Taft, C.A., de Lazaro, S.R. (eds.): Progress in Hydrogen Energy, Fuel Cells, Nano-Biotechnology and Advanced, Bioactive Compounds. Springer Nature Switzerland AG, Cham (2024). https://doi.org/10.1007/978-3-031-75984

Synthesis and Application of Silver Nanoparticles in the Development of Paints with Antimicrobial Properties

Suzane Quintana Gomes, Suelen Santos da Silva, Natalia Barbosa Preve, Jamal Rafique, Sumbal Saba, Shirley Kuhnen, Rayanne Ribeiro, Ismael Raitz, and Tiago Elias Allievi Frizon

Abstract Silver nanoparticles (AgNPs) are nanometric-scale structures widely applicable across numerous fields owing to their unique properties. Their synthesis typically has a bottom-up approach via chemical reduction reactions using agents such as sodium citrate and ascorbic acid to yield stable colloids. In this process, various factors affect the particle nucleation and growth which define their final characteristics. Sodium citrate also acts as a stabilizing agent, while ascorbic acid provides an eco-friendly and effective route for AgNP production. These nanoparticles exhibit pronounced antimicrobial activity against a wide spectrum of microorganisms, including bacteria, fungi, and viruses. This chapter presents the synthesis of AgNPs using sodium citrate and ascorbic acid, their characterization, and the results of antimicrobial assay of the nanoparticles both in solution and when incorporated into water-based paint.

Keywords Silver nanoparticles · Antimicrobial activity · Water-based paint

S. Q. Gomes (✉) · S. S. da Silva · N. B. Preve · I. Raitz · T. E. A. Frizon
Departamento de Energia e Sustentabilidade, Universidade Federal de Santa Catarina, Araranguá, Santa Catarina, Brazil

J. Rafique
Instituto de Química, Universidade Federal de Mato Grosso do Sul, Campo Grande, Mato Grosso do Sul, Brazil

S. Saba
Instituto de Química, Universidade Federal de Goiás, Goiânia, Brazil

S. Kuhnen · R. Ribeiro
Laboratório de Bioquímica e Produtos Naturais - LABINAT, Departamento de Zootecnia e Desenvolvimento Rural, Universidade Federal de Santa Catarina, Florianópolis, Brazil

C. A. Taft, S. R. d. Lazaro (eds.), *Innovative Materials for Energy and Sustainability*, Engineering Materials,
https://doi.org/10.1007/978-3-032-09149-9_6

1 Introduction

1.1 General Synthesis of Silver Nanoparticles

Nanoparticles are structures with dimensions ranging between 1 and 100 nm, and in certain biotechnological contexts, they measure up to 500 nm [1, 2]. Metallic nanoparticles have attracted particular attention due to their broad applicability in fields such as engineering, chemistry, physics, and biology. The understanding and control of their physicochemical properties, stability, and morphology—crucial factors for performance—depend on various experimental techniques, which have been extensively studied and developed over recent decades [2, 3].

This chapter addresses the synthesis of silver nanoparticles (AgNPs) using the bottom-up approach, in which nanoparticles are formed by the arrangement of atoms, molecules or small particles [4]. The most common methodology relies on chemical reduction reactions to form stable colloidal dispersions in aqueous or organic solvent media [5]. Common reducing agents include borohydride, citrate, ascorbate ions, and molecular hydrogen [6]. Scheme 1 presents the chemical equations related to the reduction of silver ions (Ag^+) to metallic silver (Ag^0) in the presence of these reducing agents.

After the reduction of silver ions (Ag^+) to metallic silver atoms (Ag^0), nucleation and subsequent agglomeration of oligomers occur to form colloidal silver particles [7]. For characterization purposes, nanomaterials often exhibit properties distinct from their micro- or macroscopic counterparts—among them, optical behaviour. In the case of AgNPs, their size being significantly smaller than the wavelength of visible light results in a characteristic yellow coloration, attributed to a strong absorption band in the range of 380–400 nm, known as surface plasmon resonance (SPR), a phenomenon which light interacts with electrons on a colloidal nanoparticles surface, causing them to oscillate collectively and creating a surface plasmon. This property enables visual monitoring of AgNP formation in solution, seen as a color shift from colorless (silver nitrate) to bright yellow, indicating nanoparticle dispersion [8].

The nucleation and growth process of AgNPs involve two interrelated steps. Initially, rapid reduction of Ag^+ to Ag^0 produce small nuclei (molecular aggregate with ~0.7 to 2 nm diameters) within milliseconds [9, 10]. This nucleation stage follows the LaMer model, driven by supersaturation and forming stable “seeds”.

$$\text{(a) } 2Ag^+_{(aq)} + 2BH_4^-{}_{(aq)} \rightarrow 2Ag_{(s)} + B_2H_{6(aq)} + H_{2(g)}$$

$$\text{(b) } 4Ag^+_{(aq)} + C_6H_5O_7^{3-}{}_{(aq)} + 2H_2O_{(\ell)} \rightarrow 4Ag_{(s)} + C_6H_5O_7H_{3(aq)} + H^+_{(aq)} + O_{2(g)}$$

$$\text{(c) } 2Ag^+_{(aq)} + C_6H_8O_{6(aq)} \rightarrow 2Ag^0_{(s)} + C_6H_6O_{6(aq)} + 2H^+_{(aq)}$$

$$\text{(d) } 2Ag^+_{(aq)} + H_{2(g)} \rightarrow 2Ag_{(s)} + 2H^+_{(aq)}$$

Scheme 1 Chemical reduction of silver ions to metallic silver using different reducing agents

Subsequently, particle growth proceeds via aggregation and coalescence, yielding larger structures (e.g., ~2 to ~12 nm within seconds to minutes) [10].

Ostwald ripening may follow, wherein smaller particles dissolve and feed the growth of larger ones, refining the size distribution [11]. In seed-mediated approaches, temporal separation between nucleation and growth allows precise control of particle shape and size [12].

Thus, the number and initial size of particles are determined during nucleation, while growth—affected by factors such as surfactants, temperature, reagent addition rate, and reductants—modulates morphology, size, and monodispersity of the AgNPs.

Among the four reducing agents previously mentioned, sodium citrate and ascorbic acid stand out due to low cost and low toxicity. Sodium citrate ($Na_3C_6H_5O_7$) plays a role as both a reducing and stabilizing agent, reducing Ag^+ to Ag^0 in hot aqueous conditions (~100 °C), while adsorbing to particle surfaces to prevent aggregation and ensure colloidal stability [13]. In biomedical applications such as antimicrobial formulations, citrate-based methods yield AgNPs averaging 30 nm with favourable stability and biocompatibility [14].

Ascorbic acid (vitamin C) is widely used for AgNP synthesis due to its simplicity, efficiency, and eco-friendliness. In aqueous solutions—typically mildly alkaline and at room temperature—ascorbic acid donates electrons to reduce Ag^+ to Ag^0, forming well-dispersed nanoparticles with SPR bands between 390 and 420 nm [15]. The concentration of ascorbic acid is crucial for determining particle size (~11 to 27 nm) and colloidal stability, with ideal conditions around 100 mM at alkaline pH for uniform size and low aggregation [15, 16]. Ascorbic acid may also stabilize AgNPs through surface interactions, preserving the integrity of the particles for weeks.

1.2 Silver Nanoparticles as Antimicrobial Agents

Silver nanoparticles (AgNPs) exhibit potent and broad-spectrum antimicrobial activity, demonstrating efficacy against Gram-positive and Gram-negative bacteria, fungi, viruses, and mycobacteria [17, 18]. Their mechanisms of action range from the release of Ag^+ ions via oxidative dissolution to the generation of reactive oxygen species (ROS), disruption of cell membranes, interference with metabolic pathways, and inhibition of DNA replication. Recent studies further indicate that AgNPs may enhance the efficacy of antibiotics by reducing minimum inhibitory concentrations (MICs) and reversing resistance in pathogens such as *Escherichia coli*, *Staphylococcus aureus*, *Acinetobacter baumannii*, and *Pseudomonas aeruginosa* [16]. Furthermore, the incorporation of AgNPs into surfaces, medical devices, and wound dressings has shown promise in preventing nosocomial infections and controlling bacterial biofilms, enabling new perspectives for clinical and public health applications [17–19].

1.3 Use of Silver Nanoparticles in Paint Formulations

Silver nanoparticles (AgNPs) have been incorporated into water-based paints to impart antimicrobial and antifouling functionalities across a variety of surfaces [19, 20]. A recent investigation assessed the performance of wood coatings containing colloidal silver, demonstrating that, beyond serving as a pigment, silver affords robust bactericidal protection without compromising the aesthetic or structural integrity of the paint. Notably, after 18 months, the coatings retained color stability ($\Delta E \approx 1.84$) and remained effective against microbial colonization. This sustained performance is attributed to the high affinity of Ag^+ ions for sulfur-containing proteins and phosphate groups present in pathogenic microorganisms [20].

Furthermore, AgNPs have been integrated into antifouling paints for applications in marine environments and construction industry. Aqueous formulations containing silver and copper exhibited significant inhibitory effects against *Bacillus* spp. under ISO-standard testing conditions, even after 12 months of exposure, achieving antimicrobial resistance (R) values exceeding the accepted efficacy threshold ($R > 2$) across multiple strains [21]. Complementary studies involving epoxy paints with AgNPs reported superior anticorrosive and antibacterial properties compared to nanoparticle-free analogues, while preserving mechanical and aesthetic performance. Collectively, these findings emphasize the versatility and practicality of AgNPs as functional additives in high-performance paints, suitable for both indoor and outdoor applications, including specialized marine antifouling systems [21, 22]. The integration of AgNPs into coating matrices offers a compelling strategy to harmonize durability, visual appeal, and antimicrobial efficacy.

2 Materials and Methods

2.1 Synthesis and Characterization of Silver Nanoparticles

Silver nanoparticles were synthesized using a procedure modified from the methodology used by Turkevich et al. [23]. Initially, 5 mL of $AgNO_3$ 1 mM (aqueous solution in ultrapure water) was added to an Erlenmeyer flask and the solution was heated to 80 °C using a heating plate under magnetic stirring. After that, 5 mL of sodium citrate 39 mM (aqueous solution) was added dropwise using a burette at a speed of 1 drop/s in approximately 5 min. The system was cooled and kept under stirring until the color transitioned from pale yellow to golden yellow (Fig. 1).

The second methodology used ascorbic acid to reduce silver nanoparticles. The synthesis started by adding 5 mL of $AgNO_3$ 1 mM (aqueous solution) to a test tube, followed by stirring and heating at 80 °C for 2 min. Then, 1 mL of ascorbic acid 0.49 mM (aqueous solution) was added dropwise until the color of the solution transitioned to a translucent yellow. At 80 °C, the formation of nanoparticles was monitored by the visible color transition from translucent yellow to golden yellow,

Fig. 1 System used for synthesizing silver nanoparticles

and by UV-VIS spectroscopy (Kasuaki IL-593-S-BI) by performing wavelength scans in the range of 800 to 190 nm. After synthesis, the silver nanoparticles were characterized by dynamic light scattering (DLS) and Zeta potential techniques. For this purpose, equipment Litesizer 500 was used with an adjusted voltage at 200 V with 100 processed runs at 22 °C.

2.2 *Incorporation of Silver Nanoparticles into Water-Based Paints*

The synthesized silver nanoparticles were incorporated into a water-based paint formulation by *Renner*® at concentrations of 0% (control), 5%, 10% and 20% (volume) with the aim of evaluating the bactericidal efficacy of the resulting paints.

2.3 Evaluation of Antimicrobial Activity

The evaluation of the antimicrobial activity of silver nanoparticles reduced with sodium citrate and ascorbic acid was performed by adapting the broth microdilution method in 96-well microplates (CLSI, 2006). Eight different concentrations (50%, 25%,12.5%, 6.25%, 3.12%, 1.56%, 0.78% and 0.39% of original solution) were tested against the standard strain of *Staphylococcus aureus* ATCC25923 in duplicate. When necessary, the products were diluted in Mueller Hinton (MH) broth. To perform the assay, the initial population density of the bacterial inoculum was standardized by controlling the turbidity equivalent to a 0.5 McFarland standard solution (1.5×10^7 to 1.5×10^8 CFU mL^{-1}). After dilution, 10 μL of the bacterial inoculum standard, corresponding to 10^5 CFU mL^{-1}, was added to the microplate wells. A positive growth control (containing inoculum), negative control 1 (absence of bacterial inoculum) and negative control 2 (broth added with different concentrations of nanoparticles in the absence of bacterial inoculum) were included in the assays. The microplate was incubated at 37 °C for 18–24 h and the minimum inhibitory concentration (MIC) was determined by the absence of microbial growth. The MIC was confirmed by the addition of 50 μL of resazurin (7-hydroxy-3*H*-phenoxazin-3-one-10-oxide) resulting in 100 μg mL^{-1} concentration. Resazurin exhibits a blue coloration in the absence of bacterial growth, as metabolically active cells are capable of reducing it to resorufin, which displays a pink color (presence of bacterial growth) [24].

The nanoparticles were also evaluated after incorporation into water-based paint at a concentration of 20%. Through this methodology, the minimum inhibitory concentration (MIC) is determined as the lowest concentration of an antimicrobial agent capable of completely inhibiting the visible growth of bacteria in the microdilution wells.

3 Results and Discussion

3.1 Development of Methodologies for Obtaining Silver Nanoparticles and Characterization

Following the implementation of two adapted methodologies for the synthesis of silver nanoparticles employing sodium citrate and ascorbic acid as reducing agents, the resulting samples were subjected to structural characterization and physicochemical property evaluation. After nanoparticle formation, all samples were stored at approximately 5 °C, and non-light exposure by wrapping the vials in aluminium foil. In Fig. 2a, it is possible to observe the golden yellow coloration expected for solutions containing silver nanoparticles synthesized with sodium citrate, just as in Fig. 2b the translucent yellow coloration was expected for silver nanoparticles using

Fig. 2 (**a**) Solution containing silver nanoparticle synthesized with sodium citrate; (**b**) solution containing silver nanoparticle synthesized with ascorbic acid

ascorbic acid. The results are promising, as silver in nanoparticle form exhibits visible coloration ranging from yellow to orange for sizes between 20 and 30 nm [25].

To confirm the formation of silver nanoparticles, AgNPs were structurally characterized by UV-Vis absorption analysis. The absorption spectra shown in Fig. 3 presents a characteristic plasmonic band of silver nanoparticles in the 424–440 nm region. Nanoparticles reduced with sodium citrate exhibited a maximum absorbance of 0.74 a.u. (Fig. 3a), while ascorbic acid presented a maximum absorbance of 0.91 a.u. (Fig. 3b). The results confirm the successful synthesis of the targeted nanoparticles.

The size of the synthesized nanoparticle and its stability were determined by dynamic light scattering (DLS) and Zeta potential, resulting in stable nanoparticles with an average size of 42 nm and a Zeta potential of −44 mV with mobility and conductivity as −32,912 μm cm/V s and 0.498 mS/cm respectively (Fig. 4) [26, 27].

3.2 *Incorporation of Silver Nanoparticles into Water-Based Paint*

A solution containing the AgNPs previously synthesized and characterized was incorporated into a water-based paint. For this purpose, different quantities of silver nanoparticles were added to the water-based paint, with the aim of evaluating the efficiency of the bactericidal activity. Assays were also carried out with different

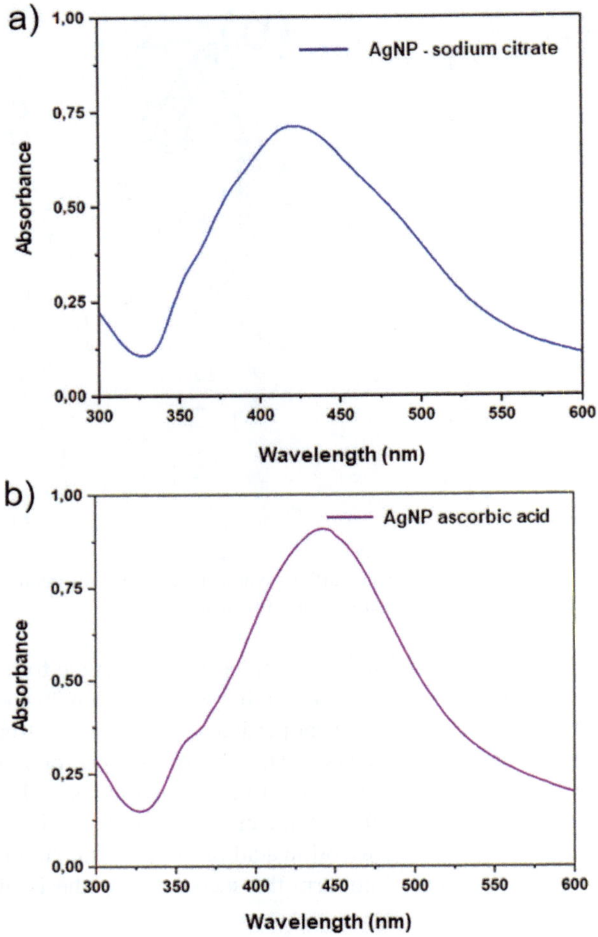

Fig. 3 UV-Vis absorption spectra for solutions containing silver nanoparticles: (**a**) reduced by sodium citrate; (**b**) reduced by ascorbic acid

resins and paints based on organic solvents; however, stability issues were observed such as inhomogeneous dispersion, attributed to the difficulty in ensuring the solubility and proper dispersion of the nanoparticles in the paint matrix. Therefore, the antimicrobial activity analyses were performed only for water-based paints.

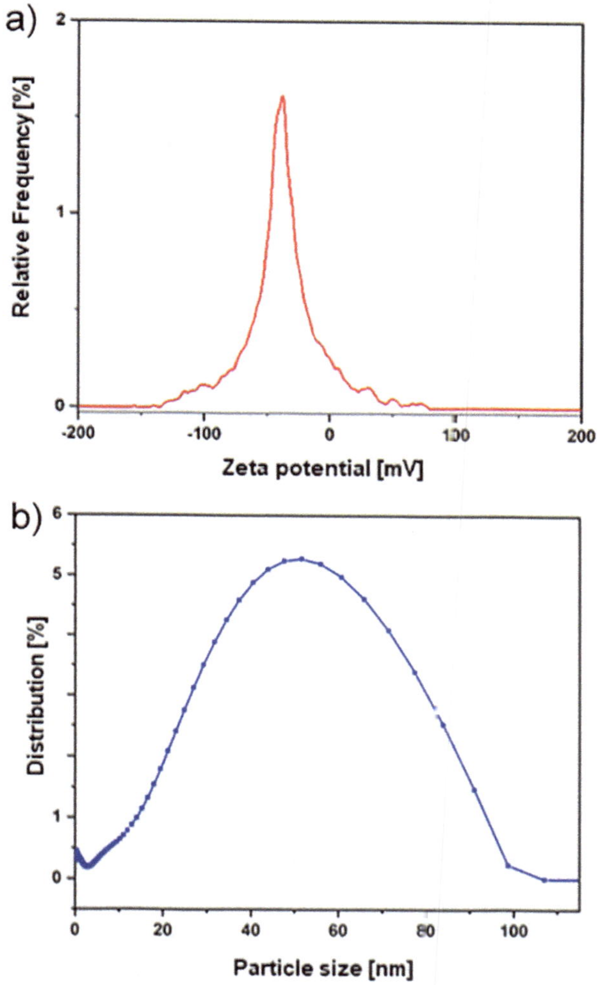

Fig. 4 (**a**) Nanoparticle stability assessed via Zeta potential. (**b**) particle size distribution measured by dynamic light scattering (DLS)

3.3 Antimicrobial Activity of Silver Nanoparticles

Two distinct experimental assays were performed to assess the antimicrobial efficacy of the synthesized silver nanoparticles: (a) using the silver nanoparticle solution directly in the wells of the microplates containing the *S. aureus* ATCC25923; (b) using the water-based paint containing silver nanoparticles applied directly to the wells of the microplates containing the *S. aureus* ATCC25923.

The antimicrobial activity of silver nanoparticles reduced with sodium citrate and ascorbic acid is shown in Figs. 5 and 6, respectively. Both Figs. 5 and 6 shows

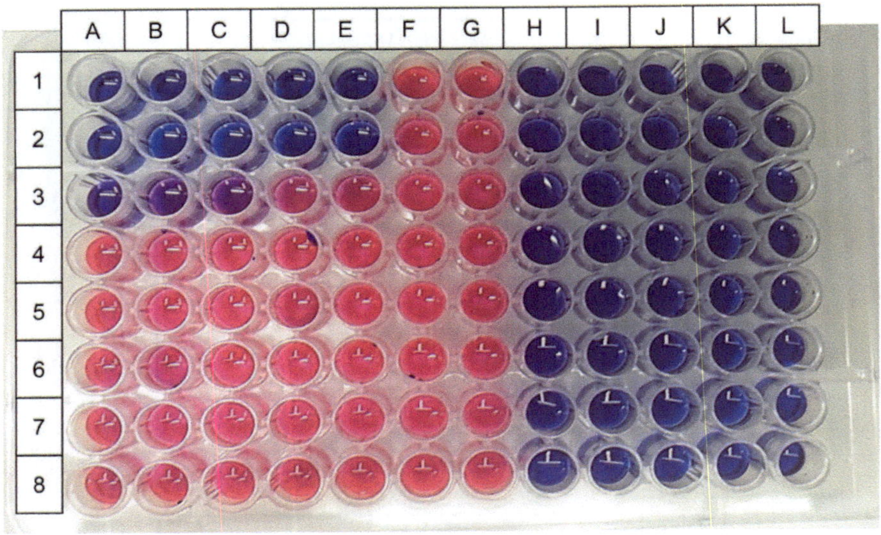

Fig. 5 Microplate stained with resazurin dye indicating the absence (blue color) and presence (pink color) of bacterial growth after exposure to different concentrations of silver nanoparticles reduced with ascorbic acid. The minimum inhibitory concentration (MIC) is indicated at 25% (2A to 2E)

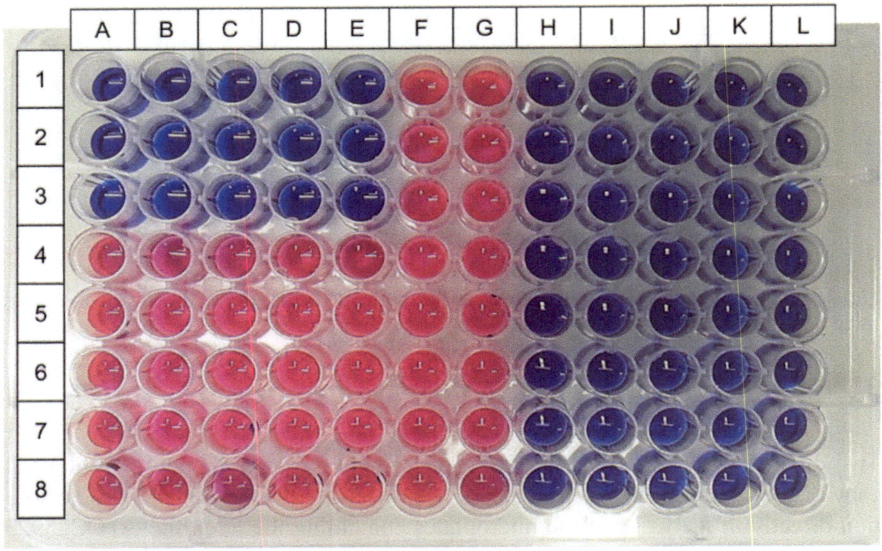

Fig. 6 Microplate stained with resazurin dye indicating the absence (blue color) and presence (pink color) of bacterial growth after exposure to different concentrations of silver nanoparticles reduced with sodium citrate. The minimum inhibitory concentration (MIC) is indicated at 12.5% (3A to 3E)

the microplate stained with resazurin dye indicating the absence (blue color) and presence (pink color) of bacterial growth after exposure to different concentrations of silver nanoparticles.

Microplates from A to E correspond to bacterial inoculum plus AgNPs solution, positive growth control were arranged on F and G, negative control 1 were arranged from H to J, and negative control 2 were arranged on K and L microplates. Microplates numbered from 1 to 8 correspond to the minimum inhibitory concentration (MIC) from 50% to 0.39% (see Sect. 2).

According to Fig. 5, MIC is indicated at 25% (2A–2E), since the microplates remained blue, indicating the absence of *S. aureus* and the ability of AgNPs to inhibit microbial growth in this value range. The next line (3A–3E), related to the MIC of 12.5%, shows a pink coloration in some microplates as a response to microbial growth, reiterating that 25% is the correct MIC.

As shown in Fig. 6, silver nanoparticles reduced with sodium citrate showed greater potential against the *S. aureus*, demonstrated by a lower MIC of 12.5% (v/v), in comparison to nanoparticles reduced with ascorbic acid with MIC of 25% (v/v) (Table 1). Furthermore, the silver nanoparticles synthesized using sodium citrate as the reducing agent were stored for 6 months post-synthesis and subsequently re-evaluated for their antimicrobial activity. In this case, the same MIC as the freshly prepared formulation was found, i.e., 12.5% (v/v).

Subsequently, silver nanoparticles were incorporated into a water-based paint following the evaluation of their antimicrobial activity. The concentration of silver nanoparticles added to the water-based paint were 0% (control), 5%, 10% and 20% (volume) to assess their effectiveness in enhancing the bactericidal activity of the resulting formulations. The control experiment containing only the water-based paint showed an antimicrobial activity of 12.5% MIC (v/v) (Table 2).

However, a reduction of the minimum inhibitory concentration (MIC) from 12.5% to 5% could be observed due to the presence of 20% (volume) of silver nanoparticles in the paint formulations (Table 2). In this case, the source of silver nanoparticles—the reducing agent employed for AgNP synthesis—had no discernible impact on the results.

Table 1 Minimum inhibitory concentration (MIC) of silver nanoparticles (AgNPs) reduced with sodium citrate or ascorbic acid against *Staphylococcus aureus* (ATCC25923)

Silver nanoparticles	MIC (v/v) (%)
AgNPs sodium citrate	12.5
AgNPs ascorbic acid	25

Table 2 Minimum inhibitory concentration (MIC) of silver nanoparticles (AgNPs) reduced with sodium citrate or ascorbic acid and incorporated in water-based paint against *Staphylococcus aureus* (ATCC25923)

Silver nanoparticles in water-based paint	MIC (v/v) (%)
AgNPs sodium citrate 20%	5
AgNPs ascorbic acid 20%	5
Control—water-based paint	12.5

4 Conclusion

Silver nanoparticles (AgNPs) synthesized using sodium citrate and ascorbic acid as reducing agents exhibited stable physicochemical characteristics and promising antimicrobial properties. Particle formation was confirmed by UV-Vis spectroscopy, which revealed the characteristic plasmon resonance band in the 424–440 nm range. Dynamic light scattering (DLS) analysis indicated an average particle size of 42 nm and a Zeta potential of −44 mV. In assays against *Staphylococcus aureus*, AgNPs reduced with sodium citrate demonstrated superior antimicrobial efficacy, exhibiting a minimum inhibitory concentration (MIC) of 12.5%, in comparison to 25% for those synthesized with ascorbic acid. Upon incorporation of AgNPs into a water-based paint, a marked enhancement in bactericidal activity was observed, with the MIC decreasing to 5% in both formulations, thereby highlighting the potential of these nanoparticles for practical applications in antimicrobial materials.

References

1. Mody, V., Siwale, R., Singh, A., Mody, H.: Introduction to metallic nanoparticles. J. Pharm. Bioallied Sci. **2**, 282 (2010). https://doi.org/10.4103/0975-7406.72127
2. Jamkhande, P.G., Ghule, N.W., Bamer, A.H., Kalaskar, M.G.: Metal nanoparticles synthesis: an overview on methods of preparation, advantages and disadvantages, and applications. J. Drug Deliv. Sci. Technol. **53**, 101174 (2019). https://doi.org/10.1016/j.jddst.2019.101174
3. Vijayakumar, M., Priya, K., Nancy, F.T., Noorlidah, A., Ahmed, A.B.A.: Biosynthesis, characterisation and anti-bacterial effect of plant-mediated silver nanoparticles using Artemisia nilagirica. Ind. Crop. Prod. **41**, 235–240 (2013). https://doi.org/10.1016/j.indcrop.2012.04.017
4. Mukherjee, P., Ahmad, A., Mandal, D., Senapati, S., Sainkar, S.R., Khan, M.I., Parishcha, R., Ajaykumar, P.V., Alam, M., Kumar, R., Sastry, M.: Fungus-mediated synthesis of silver nanoparticles and their immobilization in the mycelial matrix: a novel biological approach to nanoparticle synthesis. Nano Lett. **1**, 515–519 (2001). https://doi.org/10.1021/nl0155274
5. Wiley, B., Sun, Y., Mayers, B., Xia, Y.: Shape-controlled synthesis of metal nanostructures: the case of silver. Chem. Eur. J. **11**, 454–463 (2005). https://doi.org/10.1002/chem.200400927
6. Sharma, V.K., Yngard, R.A., Lin, Y.: Silver nanoparticles: green synthesis and their antimicrobial activities. Adv. Colloid Interf. Sci. **145**, 83–96 (2009). https://doi.org/10.1016/j.cis.2008.09.002
7. Kapoor, S., Lawless, D., Kennepohl, P., Meisel, D., Serpone, N.: Reduction and aggregation of silver ions in aqueous gelatin solutions. Langmuir. **10** (1994). https://doi.org/10.1021/la00021a026
8. Saba, S., Preve, N.B., Granja, I.J.A., Pedroso, G.J., Cabreira, C.R., Dreyer, J.P., Ribeiro, L.F.B., Horn, A.P., Marinho, M.A.G., Bellettini, I.C., Pich, C.T., Kokuszi, L.T.F., Borges, E., de Lima, V.R., Rafique, J., Frizon, T.E.A.: Synthesis of silver nanoparticles coupled with aromatic diselenides: greener approach, potential against glioma cells and DNA interaction. New J. Chem. **47**, 2727–2735 (2023). https://doi.org/10.1039/D2NJ04810H
9. Thanh, N.T.K., Maclean, N., Mahiddine, S.: Mechanisms of nucleation and growth of nanoparticles in solution. Chem. Rev. **114**, 7610–7630 (2014). https://doi.org/10.1021/cr400544s
10. Wang, F., Richards, V.N., Shields, S.P., Buhro, W.E.: Kinetics and mechanisms of aggregative nanocrystal growth. Chem. Mater. **26**, 5–21 (2014). https://doi.org/10.1021/cm402139r

11. Li, Y., Liao, Q., Hou, W., Qin, L.: Silver-based surface plasmon sensors: fabrication and applications. Int. J. Mol. Sci. **24**, 4142 (2023). https://doi.org/10.3390/ijms24044142
12. Jiang, X., Chen, W., Chen, C., Xiong, S., Yu, A.: Role of temperature in the growth of silver nanoparticles through a synergetic reduction approach. Nanoscale Res. Lett. **6**, 32 (2010). https://doi.org/10.1007/s11671-010-9780-1
13. Zhang, Q., Li, N., Goebl, J., Lu, Z., Yin, Y.: A systematic study of the synthesis of silver nanoplates: is citrate a "magic" reagent? J. Am. Chem. Soc. **133**, 18931–18939 (2011). https://doi.org/10.1021/ja2080345
14. Tareq Mahdi, M.: The reductive synthesis of silver nanoparticles using sodium citrate and their potential medical uses in skin diseases. Clin. Med. Health Res. J. **3**, 646–651 (2023). https://doi.org/10.18535/cmhrj.v3i6.265
15. Guzmán, K., Kumar, B., Grijalva, M., Debut, A., Cumbal, L.: Ascorbic acid-assisted green synthesis of silver nanoparticles: pH and stability study. In: Green Chemistry—New Perspectives. IntechOpen, London (2022)
16. Janah, I.M., Roto, R., Siswanta, D.: Effect of ascorbic acid concentration on the stability of tartrate-capped silver nanoparticles. Indones. J. Chem. **22**, 857 (2022). https://doi.org/10.22146/ijc.73196
17. Rodrigues, A.S., Batista, J.G.S., Rodrigues, M.Á.V., Thipe, V.C., Minarini, L.A.R., Lopes, P.S., Lugão, A.B.: Advances in silver nanoparticles: a comprehensive review on their potential as antimicrobial agents and their mechanisms of action elucidated by proteomics. Front. Microbiol. **15**, 1440065 (2024). https://doi.org/10.3389/fmicb.2024.1440065
18. Barua, N., Buragohain, A.K.: Therapeutic potential of silver nanoparticles (AgNPs) as an antimycobacterial agent: a comprehensive review. Antibiotics. **13**, 1106 (2024). https://doi.org/10.3390/antibiotics13111106
19. Chegini, Z., Shariati, A., Alikhani, M.Y., Safaiee, M., Rajaeih, S., Arabestani, M., Azizi, M.: Antibacterial and antibiofilm activity of silver nanoparticles stabilized with C-phycocyanin against drug-resistant Pseudomonas aeruginosa and Staphylococcus aureus. Front. Bioeng. Biotechnol. **12**, 1455385 (2024). https://doi.org/10.3389/fbioe.2024.1455385
20. Calovi, M., Coroneo, V., Rossi, S.: Antibacterial efficiency over time and barrier properties of wood coatings with colloidal silver. Appl. Microbiol. Biotechnol. **107**, 5975–5986 (2023). https://doi.org/10.1007/s00253-023-12710-1
21. Loredo-Becerra, G.M., Durán-Almendárez, A., Calvillo-Anguiano, A.K., DeAlba-Montero, I., Hernández-Arteaga, L.O., Ruiz, F.: Waterborne antifouling paints containing nanometric copper and silver against marine *Bacillus* species. Bioinorg. Chem. Appl. **2022**, 2435756 (2022). https://doi.org/10.1155/2022/2435756
22. Shi, X., Liang, H., Li, Y.: Review of progress in marine anti-fouling coatings: manufacturing techniques and copper- and silver-doped antifouling coatings. Coatings. **14**, 1454 (2024). https://doi.org/10.3390/coatings14111454
23. Turkevich, J., Stevenson, P.C., Hillier, J.: A study of the nucleation and growth processes in the synthesis of colloidal gold. Discuss. Faraday Soc. **11**, 55 (1951). https://doi.org/10.1039/df9511100055
24. Sarker, S.D., Nahar, L., Kumarasamy, Y.: Microtitre plate-based antibacterial assay incorporating resazurin as an indicator of cell growth, and its application in the in vitro antibacterial screening of phytochemicals. Methods. **42**, 321–324 (2007). https://doi.org/10.1016/j.ymeth.2007.01.006
25. Mulfinger, L., Solomon, S.D., Bahadory, M., Jeyarajasingam, A.V., Rutkowsky, S.A., Boritz, C.: Synthesis and study of silver nanoparticles. J. Chem. Educ. **84**, 322 (2007). https://doi.org/10.1021/ed084p322
26. Bhattacharjee, S.: DLS and zeta potential—what they are and what they are not? J. Control. Release. **235**, 337–351 (2016). https://doi.org/10.1016/j.jconrel.2016.06.017
27. Taft, C.A., de Lazaro, S.R. (eds.): Progress in Hydrogen Energy, Fuel Cells, Nano-Biotechnology and Advanced, Bioactive Compounds. Springer Nature Switzerland AG, Cham (2024). https://doi.org/10.1007/978-3-031-75984

Sustainable Alternatives for the Beneficiation of Pyrite from Coal Mining Waste Envisioning Future Technological Applications

Suzane Quintana Gomes, Adilson Américo, Luiz Fernando Belchior Ribeiro, Fernando Ely, Elidio Angioletto, Eduardo Junca, Ismael Raitz, and Tiago Elias Allievi Frizon

Abstract This study presents a sustainable method to purify pyrite (byproduct of coal mining) increasing its purity from less than 10% to approximately 80% using an aqueous solution of lithium heteropolytungstate (LST), a non-toxic substance that enables separation through simple sedimentation processes. The purified pyrite was then used, in combination with polymers—polystyrene (PS) and polyacrylonitrile (PAN)—for the development of novel composite filaments by electrospinning and wet-spinning techniques.

Keywords Pyrite · Sustainability · Composite filaments · Electrospinning · Wet-spinning

S. Q. Gomes (✉) · A. Américo · L. F. B. Ribeiro · I. Raitz · T. E. A. Frizon
Departamento de Energia e Sustentabilidade, Universidade Federal de Santa Catarina, Araranguá, Santa Catarina, Brazil

F. Ely
Centro de Tecnologia da Informação Renato Archer, Campinas, São Paulo, Brazil

E. Angioletto
Universidade do Extremo Sul Catarinense, Criciúma, Santa Catarina, Brazil

E. Junca
Programa de Pós-Graduação em Ciência e Engenharia de Materiais (PPGCEM), Universidade do Extremo Sul Catarinense (UNESC), Criciúma, Santa Catarina, Brazil

C. A. Taft, S. R. d. Lazaro (eds.), *Innovative Materials for Energy and Sustainability*, Engineering Materials,
https://doi.org/10.1007/978-3-032-09149-9_7

1 Introduction

1.1 *Technological Perspectives on Coal Waste Repurposing*

Renewable energy sources such as wind, hydroelectric, solar, and biomass are recognized as the most environmentally promising methods of energy generation. However, the persistent increase in global energy demands suggests that the energy matrix may continue to rely also on non-renewable sources. Given this scenario, the application of residual materials derived from coal mining—specifically from coal extracted for energy generation—becomes an opportunity to add value to mining waste while supporting the sustainability of the production chain [1–5].

1.2 *Beneficiation and Utilization of Upgraded Pyrite*

To align coal extraction with environmentally sustainable development, it is imperative to address not only the commercial coal product but also the inorganic byproducts. Reclassifying these so-called "waste" minerals as raw materials for various chemical processes represents a paradigm shift. In particular, waste streams rich in sulphide minerals exacerbate acid mine drainage (AMD), rendering water sources contaminated with bioaccumulation of metals unfit for domestic or agricultural use [6].

Sterile waste (or waste rocks) generated from coal mining exhibits elevated concentrations of metallic elements, and it is rich in iron sulfide (FeS_2). Upon exposure to water and oxygen, these materials initiate acidification processes in a cascade of subsequent chemical reactions shown in the following equations:

$$2FeS_{2(s)} + 7O_{2(g)} + 2H_2O_{(l)} \rightarrow 2Fe^{2+}{}_{(aq)} + 4SO_4{}^{2-}{}_{(aq)} + 4H^+{}_{(aq)} \tag{1}$$

$$4Fe^{2+}{}_{(aq)} + O_{2(g)} + 4H^+{}_{(aq)} \rightarrow 4Fe^{3+}{}_{(aq)} + 2H_2O_{(l)} \tag{2}$$

$$Fe^{3+}{}_{(aq)} + 3H_2O_{(l)} \rightarrow Fe(OH)_{3(s)} + 3H^+{}_{(aq)} \tag{3}$$

The ferrous ion (Fe^{2+}) generated in Eq. (1) can be oxidized to ferric ion (Fe^{3+}) as shown in Eq. (2), which undergoes hydrolysis producing further acidity (Eq. 3) [7]. The precipitated iron hydroxides are responsible for the reddish-orange color typical of AMD, frequently observed in streams surrounding coal basins. AMD is particularly prevalent in environments rich in sulphides such as pyrite (FeS_2) [8]. Similar challenges are noted in other mineral deposits, as mining activities generate vast amounts of waste containing sulphides. Without proper management, these materials undergo natural weathering and contribute to environmental degradation [9].

To mitigate these effects, it is essential to introduce reagents that reduce toxicity. Previous studies by Menezes et al. [10] and Cooling et al. [11] have demonstrated the feasibility of ferric coagulant production through controlled pyrite leaching. These investigations reported successful applications using concentrated pyrite fractions. Conventional mineral separation techniques applicable to pyrite beneficiation include density-based separation, leaching, and flotation [12–14]. Among these, dense media separation using heavy liquids stands out for its operational simplicity, precise cut-off accuracy, and reliability.

However, identifying a heavy liquid that meets specific density and viscosity criteria for optimal mineralogical separation is challenging. Although halogenated organic solvents such as bromoform have been used [13], their high toxicity and cost hinder industrial-scale implementation. Seeking a safer and more environmentally friendly alternative, the present study employs an aqueous solution of lithium heteropolytungstate (LST). LST is a non-toxic compound with density regulation, low viscosity, and chemical inertness—advantageous for pyrite separation.

Furthermore, the method adopted here allows the LST solution to be recycled, making it an economically and environmentally promising strategy for pyrite recovery from coal waste.

In addition, the potential of combining LST-based purification with high-value material development may be explored through the preliminary incorporation of purified pyrite into composite filaments using different spinning techniques. Initial attempts suggest that textile filaments and substrates with relatively high pyrite content and promising dispersion within polymer matrices may be achievable, while the electrical properties of these materials and the feasibility of developing smart fabrics based on these substrates are currently under investigation. This approach could potentially enabling new directions for innovation in the textile industry through the creation of high-value smart textiles.

1.3 Electrospinning

Electrospinning is a multifaceted process in which fine-tuning of parameters determines the quality and morphology of the fibers produced. Critical factors such as surface tension, viscosity, solvent properties, electrical conductivity, applied voltage, capillary dimensions, collector type, ambient humidity, and temperature all play decisive roles in shaping fiber characteristics.

Polymer solutions with low concentrations tend to form beads due to surface tension minimizing the solvent's surface area, resulting in spherical structures [15]. Upon solvent evaporation, these beads remain embedded within the polymer matrix. In other hand, excessively high concentrations may clog the capillary as the solvent rapidly evaporates at the tip, interrupting the continuous flow [16].

Thus, the solvent selection is critical—it must solubilize the polymer completely while exhibiting controlled volatility. Highly volatile solvents may evaporate too quickly and block the capillary tip, whereas low-volatility solvents may not fully evaporate during electrospinning, leading to bead formation or interconnected fibers [16]. Moreover, both the use of solvent mixtures and the solvent's conductivity significantly influence fiber morphology.

Increasing conductivity elevates charge density, facilitating the formation of a Taylor cone at lower voltages and reducing fiber diameter [15]. However, beyond a critical threshold, higher conductivity can decrease the effective electric field, hindering fiber stretching and consequently increasing fiber diameter [16]. A high-voltage power source is indispensable: increasing the applied voltage intensifies electrostatic forces, accelerating the jet and contributing to thinner fibers. Nonetheless, excessive voltage can destabilize the Taylor cone, promoting bead formation or irregular fiber growth [16].

The capillary and collector characteristics also critically affect the electrospinning outcome. For instance, the distance between the capillary and the collector influences both the electric field strength and the jet's travel time. If the gap is too small, the jet may move too quickly for complete solvent evaporation, favouring bead formation or creating interconnected fibers. Conversely, a distance that is too great weakens the electric field, reducing the jet's elongation and resulting in larger diameter fibers [15]. Collector material and geometry influence fiber deposition: conductive collectors efficiently dissipate charges, attracting more fibers, while insulating collectors may accumulate charge and create repulsive interactions. Rotating or parallel-plate collectors are ideal for aligned fiber formation. As for the capillary, its internal diameter affects the solution's flow rate and solvent evaporation. Larger diameters may facilitate evaporation but also raise clogging risks. Studies have shown varying fiber diameter behaviours depending on the polymer used: in some cases, fiber diameter increases with capillary size, while in others, the mean diameter remains stable though the size distribution broadens [15].

External factors such as ambient humidity and temperature also influence the electrospinning process. High humidity may lead to water condensation on the fiber surface, causing pore formation. Temperature, in turn, affects solvent evaporation rates and solution viscosity [17]. Lower temperatures decrease evaporation, allowing greater jet elongation before solidification, potentially yielding finer fibers. Higher temperatures reduce viscosity, which can also lead to smaller diameters—but this effect must be carefully managed to avoid undesirable results [17].

1.4 Wet Spinning

Wet spinning is a traditional fiber-forming technique in which a polymer dissolved in a solvent is extruded into a coagulation bath to form continuous fibers [18]. In this process, the polymer solution is injected into a coagulating or crosslinking medium, enabling the formation of fibers with varying morphologies, diameters (30–600 μm), and porosities (250–500 μm) [19]. Fiber diameter can be controlled by adjusting the feed rate or the diameter of the extrusion needle. This method is versatile and applicable to both natural and synthetic polymers [18, 20].

One significant advantage of wet spinning is its operation at low temperatures, which favours cell viability when applied in tissue engineering—allowing even the encapsulation of live cells into 3D scaffolds [21]. Furthermore, the final characteristics of the fibers are heavily influenced by the nature of the solvent and the conditions of the coagulation bath. Typical wet-spun fibers include rayon, acrylics, aramids, and materials such as polyvinyl alcohol (PVA), polyacrylonitrile (PAN), and ultra-high-molecular-weight polyethylene produced via gel spinning (a wet spinning variant with an air gap) [18, 20, 22].

As in electrospinning, numerous parameters influence the wet spinning process and can significantly impact the polymer coagulation kinetics and final fiber morphology. These include polymer concentration and molecular weight, solvent/non-solvent system, additives present in the solution or bath, processing temperature, and equipment configuration. For instance, the coagulation bath may contain: a poor solvent, which partially dissolves the polymer; a non-solvent, incapable of dissolving the polymer; or a solvent/non-solvent mixture composed of one solvent that fully dissolves the polymer and another that does not [23]. Polymer coagulation begins when the extruded filament comes into contact with the bath, induced by solvent exchange. Therefore, the solidification rate and final fiber morphology are dictated by the dynamics of this solvent–non-solvent interaction. Bath temperature also plays a role in modifying filament properties, as different solvent/non-solvent combinations and temperature conditions can affect the degree of phase separation, porosity, and uniformity of the resulting fibers [23].

2 Materials and Methods

2.1 Beneficiation and Characterization of Waste Pyrite from Coal Mining

The raw pyrite-bearing material (FeS_2) derived from coal extraction processes was collected in the city of Forquilhinha in the state of Santa Catarina, Brazil. After undergoing a pre-beneficiation process, the material was subjected to density-based separation using lithium heteropolytungstate (LST)—an aqueous non-toxic solution with a density of 2.85 ± 0.02 g mL^{-1}. The crystalline phases were characterized through X-Ray Diffraction (XRD), Fourier Transform Infrared Spectroscopy (FTIR), and Energy Dispersive X-ray Fluorescence (XRF) analyses.

The complete method of beneficiation and density separation using LST, and characterization are fully described in our previous work [24].

2.2 Production of Pyrite-Based Composite Filaments with Different Polymers

Two polymers were selected for composite formulation with purified pyrite: polystyrene (PS) and polyacrylonitrile (PAN). Solubility and compatibility tests used different percentages of polymer (10%, 15%, 20%, 25%) and purified pyrite (FeS_2) (0%, 10%, 20%, 25%, 30%) with dimethylformamide (DMF) until complete dissolution, stirring 750 rpm/min at room temperature.

The production of the composite filament was carried out via wet-spinning, according to the following steps: (1) Extrusion of the composite solution; (2) Precipitation in a coagulation bath containing a non-solvent for the polymer; (3) Stretching and collecting on a rotating mandrel.

Another technique used to obtain composite filaments was electrospinning. In this method, polymeric fibers were deposited onto a conductive aluminum substrate by applying a high potential difference, resulting in the formation of a membrane (non-woven fabric) with fibers ranging from a few micrometers to nanometers in diameters. The previously solubilized samples were loaded into the syringe with a flat-tipped needle and mounted on a syringe pump. Electrodes were connected to both the needle tip and the collector to establish the required electric field for the electrospinning process. The process was conducted for approximately 90 min until microfibers were obtained from the collector. Fiber morphology and surface topology were analysed via Scanning Electron Microscopy (SEM) using a Hitachi TM3030 microscope at magnifications ranging from 300× to 4000×.

3 Results and Discussion

The use of lithium heteropolytungstate (LST), an environmentally benign aqueous solution, as the separation medium significantly enhanced both the safety of handling and the ecological footprint of the pyrite beneficiation process. Characterization analyses of the raw sample indicated that pyrite (FeS_2) comprised less than 10% of the material. However, following the LST-based separation, the pyrite content increased to approximately 76%.

This marked improvement in purity enabled the repurposing of pyrite as a precursor for advanced technological applications, particularly in the production of polymeric composite filaments. To explore this potential, composite formulations using polystyrene (PS) and polyacrylonitrile (PAN) as matrices were developed with variable pyrite loadings.

Tables 1 and 2 detail the solubility and compatibility tests between pyrite and the two polymer matrices—PS and PAN—in dimethylformamide (DMF) at room temperature. These formulations evaluated polymer concentrations ranging from 10% to 25% (PS) and 4% to 9% (PAN), along with pyrite contents varying from 0% to 60%.

Table 1 Solubility and compatibility tests of composites with pyrite and polystyrene (PS)

Entry	DMF (mL)	PS/DMF concentration (%)	Pyrite (P98) concentration (%)	Proportion of P98 on the PS in g
1	2.7	10	0	0
2	2.7	10	30	0.09
3	2.55	15	0	0
4	2.55	15	30	0.135
5	2.4	20	0	0
6	2.4	20	20	0.12
7	2.25	25	10	0.075

Table 2 Solubility and compatibility tests of composites with pyrite and polyacrylonitrile (PAN)

Entry	DMF (mL)	PAN/DMF concentration (%)	Pyrite (P98) concentration	Proportion of P98 on the PAN in g
1	1.91	4	60%	0.12
2	1.91	5	50%	0.1
3	1.91	6	40%	0.08
4	1.91	7	30%	0.06
5	2.0	8	0	0
6	1.69	9	60%	0.24

Preliminary experiments employing Pyrite (P98) + Polystyrene (PS) and Pyrite (P98) + Polyacrylonitrile (PAN), in the proportions detailed in Tables 1 and 2, revealed that the P98 + PS composite exhibited superior performance in the electrospinning process, whereas the P98 + PAN composite was more suitable for the wet-spinning technique.

Figure 1 shows the electrospinning apparatus used. For this technique to stabilize the membrane and maximize the pyrite content while ensuring a homogeneous distribution throughout the medium, the solution's density was increased to counteract the sedimentation of pyrite particles, which are significantly denser than both the DMF and PS solutions, so they tend to sink during the electrospinning process. Notably, excessive voltage from the high-voltage source accelerated the ejection of solution from the syringe, leading to particle separation, while insufficient voltage led to bead formation and polymer accumulation at the needle tip.

The electrospinning technique enabled ultrafine fibers with excellent dispersion of pyrite, though the best results were obtained with solution containing a maximum of 30% pyrite in relation to the polymer matrix. Higher concentrations resulted in needle clogging and consequent interruption of the process. Moreover, the slower extrusion rates inherent to electrospinning, when compared to wet spinning, promoted early precipitation of the pyrite within the syringe. The optimal electrospinning parameters for P98 + PS composites were as follows: syringe flow rate of 1 mL h^{-1}, applied voltage of 15 kV, and a collector distance of 15 cm.

Another crucial parameter for obtaining stable composite filaments was the polymer concentration. The best results were achieved with a PS concentration of 15%.

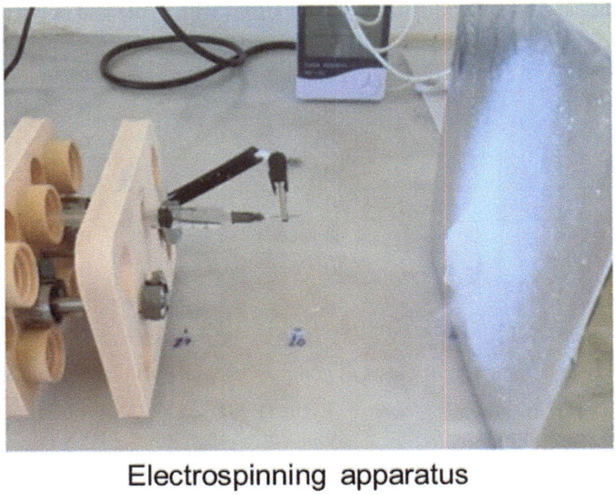

Fig. 1 Apparatus employed to produce filaments via electrospinning

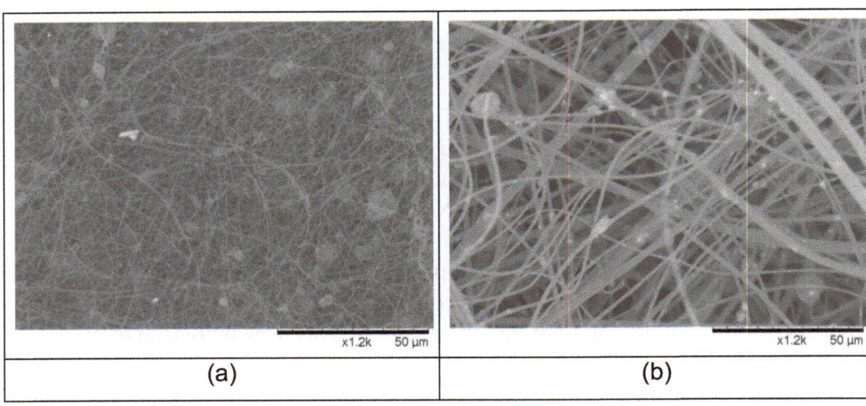

Fig. 2 Micrographs of composite fibers containing 30% pyrite with (**a**) 10% PS and (**b**) 15% PS. Magnification: 1200×

At concentrations above 20%, the solution became excessively viscous, hindering proper flow and filament formation. Conversely, at a PS concentration of 10%, nanometric fibers were produced. However, considering that the pyrite particles are a few micrometers in size, their effective incorporation into the polymer matrix was compromised under these conditions. Figure 2 compares the fibers obtained with 10% and 15% PS concentrations, both containing 30% pyrite relative to the polymer matrix. As shown in the micrograph of Fig. 2b, the pyrite particles exhibited excellent dispersion within the polymer matrix, and the viscosity adjustment also contributed to effectively reducing the formation of beads along the fibers, as observed in Fig. 2a.

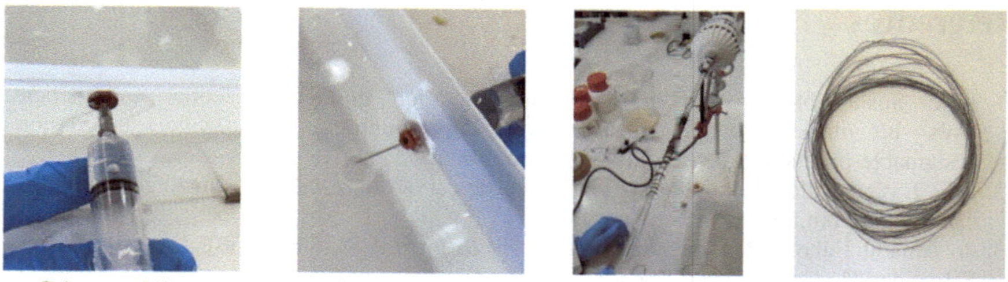

Fig. 3 Stages and apparatus used for the production of filaments via wet-spinning

For wet spinning, PAN demonstrated superior solubility and compatibility with pyrite. The composite filaments obtained via this technique accommodated pyrite concentrations ranging from 30% to 60% by mass relative to PAN. Concentrations beyond 60% led to excessive viscosity, hindering filament extrusion and interrupting the process. The solution with 60% pyrite exhibited the best performance, processed with a syringe flow rate of 0.65 mL/h. The coagulation bath consisted of a DMF/water solvent–non-solvent system. Subsequently, pyrite was integrated into the polymer solution under controlled conditions (Fig. 3).

4 Conclusion

By employing an environmentally safer, non-toxic aqueous solution of lithium heteropolytungstate (LST) for mineral beneficiation, the pyrite content in coal waste was successfully elevated from below 10% to approximately 76%. This substantial increase in purity not only minimizes environmental impacts but also enables the use of pyrite as a viable raw material for advanced technological applications—particularly in the fabrication of polymer-based composite filaments.

To that end, polystyrene (PS) and polyacrylonitrile (PAN) emerged as promising polymer matrices. Solubility and compatibility tests in DMF revealed that the pyrite–PS system was better suited for electrospinning, enabling the formation of ultrafine fiber membranes with uniform particle distribution. In contrast, the pyrite–PAN system demonstrated favourable performance in the wet spinning technique, accommodating pyrite concentrations between 30% and 60%, with 60% being the upper threshold for successful filament formation.

These findings demonstrate that, through careful adjustment of viscosity, solvent type, and solution feed rate, it is possible to incorporate substantial amounts of pyrite into polymer matrices—establishing the basis for composite filaments with potentially superior physical and functional properties.

References

1. IEA: Sustainable production of second-generation biofuels: potential and perspectives in major economies and developing countries. https://www.iea.org/reports/sustainable-production-of-second-generation-biofuels
2. SIECESC: Sindicato da indústria de extrato de carvão do estado de santa catarina. https://carvaomais.com.br/
3. do Amaral Filho, J.R., Schneider, I.A.H., de Brum, I.A.S., Sampaio, C.H., Miltzarek, G., Schneider, C.: Caracterização de um depósito de rejeitos para o gerenciamento integrado dos resíduos de mineração na região carbonífera de Santa Catarina, Brasil. REM Rev. Esc. Minas. **66**, 347–353 (2013). https://doi.org/10.1590/S0370-44672013000300012
4. Tibbett, M.: Post-mining ecosystem reconstruction. Curr. Biol. **34**, 387–393 (2024). https://doi.org/10.1016/j.cub.2024.03.065
5. Santos, C.P., Silva, J.M.M., de Oliveira, C.M., Jiusti, J., Montedo, O.R.K., Peterson, M.: Quartz quantification in porcelain stoneware tile formulations by differential scanning calorimetry. Thermochim. Acta. **705**, 179050 (2021). https://doi.org/10.1016/j.tca.2021.179050
6. Evangelou, V.P.(B.), Zhang, Y.L.: A review: pyrite oxidation mechanisms and acid mine drainage prevention. Crit. Rev. Environ. Sci. Technol. **25**, 141–199 (1995). https://doi.org/10.1080/10643389509388477
7. Xu, Y.: A study of pyrite acidification in oil sand froth treatment tailings deposits. Int. J. Environ. Sci. Nat. Resour. **27**, 556218 (2021). https://doi.org/10.19080/IJESNR.2021.27.556218
8. Qureshi, A., Maurice, C., Öhlander, B.: Potential of coal mine waste rock for generating acid mine drainage. J. Geochem. Explor. **160**, 44–54 (2016). https://doi.org/10.1016/j.gexplo.2015.10.014
9. Benzaazoua, M., Bussière, B., Kongolo, M., McLaughlin, J., Marion, P.: Environmental desulphurization of four Canadian mine tailings using froth flotation. Int. J. Miner. Process. **60**, 57–74 (2000). https://doi.org/10.1016/S0301-7516(00)00006-5
10. Menezes, J.C.C.C., Colling, A.V., Silva, R.A.S., Dos Santos, R.H., Scheneider, I.A.H.: Ferric sulphate coagulant obtained by leaching from coal tailings. Mine Water Environ. **36**, 457–460 (2017). https://doi.org/10.1007/s10230-017-0453-5
11. Colling, A.V., Menezes, J.C.S.d.S., Schneider, I.A.H.: Bioprocessing of pyrite concentrate from coal tailings for the production of the coagulant ferric sulphate. Miner. Eng. **24**, 1185–1187 (2011). https://doi.org/10.1016/j.mineng.2011.04.003
12. Machado de Oliveira, C., Gesser Müller, T., Patricio Ferreira, L., Prado Cechinel, M.A., Peterson, M., Raupp-Pereira, F.: Valorization of iron pyrite from coal mining in southern Brazil. J. Environ. Chem. Eng. **7**, 102931 (2019). https://doi.org/10.1016/j.jece.2019.102931
13. Oliveira, C.M., Machado, C.M., Duarte, G.W., Peterson, M.: Beneficiation of pyrite from coal mining. J. Clean. Prod. **139**, 821–827 (2016). https://doi.org/10.1016/j.jclepro.2016.08.124
14. Mu, Y., Peng, Y., Lauten, R.A.: The depression of pyrite in selective flotation by different reagent systems—a literature review. Miner. Eng. **96–97**, 143–156 (2016). https://doi.org/10.1016/j.mineng.2016.06.018
15. Electrospinning process. In: An Introduction to Electrospinning and Nanofibers, pp. 90–154. World Scientific, Singapore (2005)
16. Haider, A., Haider, S., Kang, I.-K.: A comprehensive review summarizing the effect of electrospinning parameters and potential applications of nanofibers in biomedical and biotechnology. Arab. J. Chem. **11**, 1165–1188 (2018). https://doi.org/10.1016/j.arabjc.2015.11.015
17. De Vrieze, S., Van Camp, T., Nelvig, A., Hagström, B., Westbroek, P., De Clerck, K.: The effect of temperature and humidity on electrospinning. J. Mater. Sci. **44**, 1357–1362 (2009). https://doi.org/10.1007/s10853-008-3010-6
18. Ozipek, B., Karakas, H.: Wet spinning of synthetic polymer fibers. In: Advances in Filament Yarn Spinning of Textiles and Polymers, pp. 174–186. Elsevier, Amsterdam (2014)

19. Mejía Suaza, M.L., Hurtado Henao, Y., Moncada Acevedo, M.E.: Wet electrospinning and its applications: a review. TecnoLógicas. **25**, e2223 (2022). https://doi.org/10.22430/22565337.2223
20. Shaikh, T., Chaudhari, S., Varma, A.: Viscose rayon: a legendary development in the manmade textile. Int. J. Eng. Res. Appl. **2**, 675–680 (2012)
21. Puppi, D., Chiellini, F.: Wet-spinning of biomedical polymers: from single-fibre production to additive manufacturing of three-dimensional scaffolds. Polym. Int. **66**, 1690–1696 (2017). https://doi.org/10.1002/pi.5332
22. Liu, S., Tan, L., Pan, D., Chen, Y.: Gel spinning of polyacrylonitrile fibers with medium molecular weight. Polym. Int. **60**, 453–457 (2011). https://doi.org/10.1002/pi.2968
23. Rohani Shirvan, A., Nouri, A., Sutti, A.: A perspective on the wet spinning process and its advancements in biomedical sciences. Eur. Polym. J. **181**, 111681 (2022). https://doi.org/10.1016/j.eurpolymj.2022.111681
24. Ghedin, S.C.D., Pedroso, G.J., Neto, J.C.B., Preve, N.B., Gondoreck, G.G., Ely, F., Angioletto, E., Ourique, F., Ribeiro, L.F.B., Frizon, T.E.A.: Processing of pyrite derived from coal mining waste by density separation technique using lithium heteropolytungstate (LST). Matéria (Rio de Janeiro). **27** (2022). https://doi.org/10.1590/1517-7076-rmat-2022-0169

MIX
Papier aus verantwortungsvollen Quellen
Paper from responsible sources
FSC® C105338

If you have any concerns about our products,
you can contact us on
ProductSafety@springernature.com

In case Publisher is established outside the EU,
the EU authorized representative is:
Springer Nature Customer Service Center GmbH
Europaplatz 3, 69115 Heidelberg, Germany

Printed by Libri Plureos GmbH
in Hamburg, Germany